Retail change

Retail change

contemporary issues

EDITED BY
Rosemary D. F. Bromley
Colin J. Thomas
University College of Swansea

First published in 1993 by UCL Press

UCL Press Limited
University College London
Gower Street
London WC1E 6BT

ISBN:
1-85728-059-8 HB
1-85728-060-1 PB

British Library Cataloguing-in-Publication Data
A catalogue record for this book
is available from the British Library.

Front cover illustration:
Royal Victoria Palace, Tunbridge Wells, Kent, England
MEPC's 300,000 sq ft retail flagship, opened April 1992.
Reproduced by kind permission of MEPC plc.

Typeset in Palatino.
Printed and bound by
Biddles Ltd, King's Lynn and Guildford, England.

CONTENTS

Preface ix
List of Contributors xi

PART ONE The changing organization of retailing 1

1 *Retail change and the issues* 2
ROSEMARY D. F. BROMLEY & COLIN J. THOMAS
The context of retail change 3
The spatial transformation of retail structure 6
The contents of the book 9

2 *The internationalization of retailing* 15
JOHN A. DAWSON
Sourcing in an international context 17
International retail operations 26
The internationalization of management ideas 36
Conclusions 39

3 *Retail concentration and the internationalization of British grocery retailing* 41
NEIL WRIGLEY
The "golden age" of British grocery retailing 43
Internationalization as a response to changing competitive conditions 49
Two facets of the internationalization of British grocery retailing 54
Conclusions 66

PART TWO Transformation and the urban region 69

4 *The proliferation of the planned shopping centre* 70
JONATHAN REYNOLDS
The definition of the planned shopping centre 71
Penetration of the planned shopping centre in Europe 73
The operational environment 82
The planning and institutional environment 84
Conclusions 86

5 *Transformation and the city centre* 88
CLIFFORD M. GUY & J. DENNIS LORD
The two city centres in the early 1970s 90
Changes in retail size and composition in the 1970s and 1980s 93
Retail development and redevelopment in the 1970s and 1980s 98
Local government support for the city centre 104
Conclusions 107

PART THREE Planning implications of retail change 109

6 *Prospects for the central business district* 110
GWYN ROWLEY
The inner city setting 110
The traditional CBD 111
The changing CBD 113
New growth centres and the CBD 115
The planning response 121
Concluding comments 125

7 *The impact of out-of-centre retailing* 126
COLIN J. THOMAS & ROSEMARY D. F. BROMLEY
The impact of superstores and hypermarkets 126
The impact of retail warehouses and retail warehouse parks 134
The impact of regional shopping centres 140

Reaction in the town centre 145
International experience 147
Conclusions 150

8 *Planning and shopper security* 153
TANER OC & SYLVIA TRENCH
Design and development solutions 156
Housing and mixed-use development 157
Dense and compact development 161
Positive identity 164
Access and urban design 165
Conclusions 167

PART FOUR Social issues of retail change 171

9 *The disadvantaged consumer: problems and policies* 172
TIM WESTLAKE
Disadvantaged consumers 174
Solutions to the problems of disadvantaged consumers 179
New alternatives 182
Conclusions 190

10 *Working conditions and the trading week* 192
DAVID A. KIRBY
Trends in the workforce 195
A questionnaire survey of retail employees 196
Research findings from the survey 197
Conclusions 205

11 *Shopping as leisure* 208
PETER NEWBY
The growth in leisure 209
The use of leisure time 210
Shopping behaviours 212

The circumstances of leisure shopping 214
The response of the retail system to leisure shopping 215
The prospects for leisure shopping 224
Conclusions 228

12 *The Greening of shopping* 229
FRANK W. HARRIS & LARRY G. O'BRIEN
Perspectives on environmental health 230
Definitions of "Green" 232
The emergence of Green retailing in the UK 232
Industry responses to environmental concern 234
Consumer responses to environmental concern 239
Conclusions 245

13 *Retail change: retrospect and prospect* 247
ROSEMARY D. F. BROMLEY & COLIN J. THOMAS
Changing retail organization 248
The new retail phenomena 249
Planning implications of retail decentralization 250
Issues of social concern 255
Social trends in retailing 256
Concluding comments 258

References 259
Index 279

PREFACE

The retail environment of developed economies has undergone radical change. Retail companies have grown at an unprecedented scale and are increasingly becoming international corporations. Consumers have become more affluent and mobile, creating demands for ever more specialized goods and services from outlets in car-orientated locations. New retail forms have emerged, reflecting a response to these demands and to the changing requirements of retailers in the evolving business, fiscal and planning environment. The past twenty-five years have witnessed the emergence of the superstore, the retail warehouse, retail parks and the increasingly ubiquitous regional shopping centre. Planning has tried to direct the development of the new facilities, while attempting to minimize their negative impacts on the traditional retail hierarchy and, in particular, on the city centre. Concern for shopper safety is also exerting additional pressure on planners to design secure environments. The problems of the disadvantaged consumer in the context of the changing distribution of retail facilities; pressures on the workforce caused by changing working conditions and the extension of the trading week; and consumer demands for a "Green" dimension in contemporary marketing strategies have all emerged as key social issues in the process of change. The increasingly explicit definition of shopping as a leisure activity is also presenting new challenges for the retail sector. These issues are the focus of this book, which aims to provide a contemporary review of organizational, locational, planning and social perspectives on the process of retail change.

The stimulus for this book was the Institute of British Geographers' conference held in Swansea in 1992. Seven of the thirteen chapters are based on research originally presented at that conference. The others were specifically commissioned to cover the key issues of particular concern to geography, planning and other disciplines embracing retail change. All the core chapters contain original research and case study material apposite to the issues in question.

Our principal acknowledgements must go to the contributors for agreeing to participate in this volume and for their willingness to undertake the suggested revisions. Tragically, one of the authors, Larry O'Brien, lost his battle against a long illness shortly after his jointly authored chapter was completed. We are particularly grateful to Guy Lewis and Nicola Jones of the Cartographic Unit at the Department of Geography, University College of Swansea for redrawing the majority of the illustrations.

Finally, we would like to dedicate this book to our children, respectively; Martin and Paul; and David.

ROSEMARY D. F. BROMLEY & COLIN J. THOMAS Swansea, January 1993

CONTRIBUTORS

Rosemary D. F. Bromley Lecturer in Geography, Department of Geography, University College of Swansea, Singleton Park, Swansea SA2 8PP, Wales

John A. Dawson Professor of Marketing, Department of Business Studies, University of Edinburgh, Edinburgh EH8 9JY, Scotland

Clifford M. Guy Senior Lecturer, Department of City and Regional Planning, University of Wales, College of Cardiff, Aberconway Building, Colum Drive, Cardiff CF1 3YN, Wales

Frank W. Harris Lecturer in Geography, Department of Geography and Geology, Cheltenham and Gloucester College of Higher Education, PO Box 220, The Park, Cheltenham, Gloucestershire GL50 2QF, England

David A. Kirby Booker Professor in Entrepreneurship, Centre for Entrepreneurship in the Service Sector, Durham University Business School, Mill Hill Lane, Durham City DH1 3LB, England

J. Dennis Lord Professor of Geography, Department of Geography and Earth Sciences, University of North Carolina at Charlotte, Charlotte, North Carolina 28223, USA

Peter Newby Head of Enterprise, Enterprise Programme, Middlesex University, Queensway, Enfield, Middlesex EN3 4SF, England

Larry G. O'Brien (Died November 1992) Formerly Lecturer in Geography, Department of Geography, University of Newcastle-upon-Tyne, England

Taner Oc Senior Lecturer, Institute of Planning Studies, University of Nottingham, University Park, Nottingham NG7 2RD, England

Jonathan Reynolds Fellow in Retail Marketing, Oxford Institute of Retail Management, Templeton College, Kennington, Oxford OX1 5NY, England

Gwyn Rowley Senior Lecturer in Geography, Department of Geography, University of Sheffield, Western Bank, Sheffield S10 2TN, England

Colin J. Thomas Lecturer in Geography, Department of Geography, University College of Swansea, Singleton Park, Swansea SA2 8PP, Wales

Sylvia Trench Lecturer in Economics and Transport Policy, Institute of Planning Studies, University of Nottingham, University Park, Nottingham NG7 2RD, England

Tim Westlake Senior Lecturer in Planning, School of Planning, Faculty of the Built Environment, University of Central England, Perry Barr, Birmingham B42 2SU, England

Neil Wrigley Professor of Geography, Department of Geography, University of Southampton, Highfield, Southampton SO9 5NH, England

PART ONE

THE CHANGING ORGANIZATION OF RETAILING

CHAPTER ONE
Retail change and the issues

Rosemary D. F. Bromley & Colin J. Thomas

Since the early 1970s there has been dramatic change in the retail environment of Western economies. This is of critical importance to both economies and societies, given the growing significance of the retail sector associated with the shift from an industrial to a post-industrial stage. In the UK, 2.1 million people (almost 10 per cent of the labour force) were employed in retail distribution in 1992 (*Employment Gazette* November 1992) and the largest retailers are now among the largest companies (Lowe & Crewe 1991). In 1991, J. Sainsbury employed about 95,000 people (Sainsbury 1991), while Marks & Spencer employed about 61,000 (Marks & Spencer 1991). In the USA, even larger organizations are commonplace. Wal-Mart, a discount retail chain, was in 1992 reputed to be the largest retailer, with a turnover of $44 billion from 1,700 stores and a workforce in excess of 365,000 (*Financial Times*, 3 March 1992).

Moreover, retailing exerts considerable influence on the morphology and functioning of Western cities. Retail activities constitute important focal points in the urban fabric, while the emergence of linear and isolated retail developments along major roads are equally significant. The contemporary city is to a substantial degree articulated in relation to retail facilities, and this has important consequences for the nature of city growth and associated opportunities and constraints for urban planning. Shopping trips are also a major component of spatial behaviour in the city, since the vast majority of the population is involved in some direct or indirect way with shopping activities. For most practical purposes, shopping is necessary to obtain the great majority of the goods and services required by modern households. For some, shopping also forms a necessary part of social interaction, while the concept of shopping as an increasingly leisure-based activity has gained widespread acceptance as an important element of popular culture.

The context of retail change

Retail change has occurred in the context of wide-ranging socioeconomic trends. First, increasing affluence has been associated with a rise in car ownership and much greater mobility. Given the improvement in roads, people are now able and willing to travel far greater distances for their shopping and the car is increasingly being regarded by retailers as a "shopping basket on wheels". However, those who do not own cars have become polarized as a disadvantaged group whose poor mobility constrains their access to urban facilities (Bromley & Thomas 1993). The growth in car ownership and the consequent volume of traffic has also had detrimental effects on movement within the city, and growing central congestion has contributed to decentralization.

A second group of trends springs from considerable changes in the spatial redistribution and composition of population. Counter-urbanization, involving the shift towards a less-concentrated pattern of population distribution, in which rural areas and small towns grow faster than large cities, was a dominant feature of the 1970s and 1980s (Champion 1989, 1992). Consumer services such as retailing have followed the population in decentralizing from city centres (Champion & Townsend 1990). Trends in composition have been equally significant. The population is ageing and the proportion of elderly people has increased markedly (Warnes 1989). In 1989, 5 per cent of males and 8 per cent of females were aged 75 or older compared with 3 per cent and 6 per cent, respectively, in 1977 (OPCS 1991: 8). Households have shrunk in size, increased in number and become more varied in type (Champion & Townsend 1990). The proportion of one-person households has grown from 17 per cent in 1971 to 25 per cent in 1989 (OPCS 1991: 11), a trend in part associated with the increase in the elderly. There are now more households to participate in shopping and more of them are experiencing the restricted mobility associated with the disabilities of age.

A third set of socioeconomic trends result from the changing character of the working population. Part-time employment for both sexes grew considerably in the 1970s and 1980s, and retailing has been prominent in this trend (Townsend 1986, Watson 1992). The proportion of women in the workforce has also increased, and retailing itself shows continued feminization of the workforce, with women comprising about

62 per cent of workers (Sparks 1992). Those in work experience time constraints on their shopping patterns, a factor of particular relevance to women, whose traditional gender rôle has included being the principal shopper (Davies & Bell 1991). However, alongside these trends the growth of unemployment since the 1980s associated with deindustrialization should be noted, and the suggested emergence of an underclass (Eversley 1990). The shopping constraints of disadvantage are a hallmark of the unemployed.

Finally, it is important to be aware of changing social and political attitudes, a topic pursued in Part 4 of this book. Leisure time has increased and people are more conscious of their use of time (Blacksell 1991). For activities such as shopping, the experience either tends towards the leisurely and pleasurable, or the quick and time-efficient. Either way, it is undeniable that retail change is having an impact upon contemporary social interaction. In North America, Hopkins (1991), for example, in a redefinition of "malingering", introduces the "mallie" as the adolescent habitué of the mall environment. This concept has its counterpart in the family-forming and middle-aged "sheddies" who focus their home-furnishing activities on the retail warehouse concentrations (Brown 1989). Similarly, the British superstore grocery shopper has almost become a cliché of middle-class family normality. During the 1980s environmental concern also emerged as a powerful force, and impinged on many areas of social and economic activity, including retailing. A significant minority of consumers are concerned about the potential global environmental impacts of their patterns of consumption. Issues such as the manufacture and use of artificial fertilizers and their associated pollutant and health effects are relevant in this respect. The burgeoning use of seemingly wasteful packaging has been of similar consequence. Likewise, the origins of imported goods have also generated concerns about possibly exploitative economic and political relationships in the source countries (Gardiner & Sheppard 1989).

While the four groups of socioeconomic trends identified above exert a background influence on retail change, other trends have a more integral rôle. These include technological change in retail operations, the growing size and internationalization of business organizations, marketing changes, and the weakening of planning constraints.

The technological revolution of the past few decades has transformed retail operations. There has been a change from counter service to self-service, and a huge growth in the size of many types of outlet, particularly in groceries and DIY. Increasingly efficient methods of handling stock and organizing distribution systems have been associated with growing economies of scale. Computerized information technology widely introduced in the 1980s has transformed stock control, distribution management and financial control systems (Wrigley 1988). Electronic point-of-sale (EPOS) data capture is now common, providing information for the quick and accurate auditing of sales and stock movements, and saving much staff time in stock control (Johnson 1987a). The EPOS technology can form the basis for sophisticated management information systems, monitoring and directing activities at the sales areas and in the warehouse and distribution networks. Other uses of information technology in retailing include private viewdata systems and remote shopping (Guy 1988a). Although this book does not focus on technological changes, Chapter 9 gives specific attention to electronic home shopping.

The business organization of retailing has also changed dramatically and the associated trends receive particular attention in Part 1 of this book. The small shop has declined at the expense of the large and changes in ownership patterns have led to the emergence of major retail corporations. There has also been a significant transformation in retail marketing and in marketing strategies since the 1960s. The abolition of retail price maintenance in 1964 in the UK, the spread of retailer own-brands and increased concentration have been associated with the emergence of retailers with greater power and influence than their manufacturing suppliers (McGoldrick 1990, Ford 1991). Retail concentration has therefore emerged as an area of key importance that is itself undergoing considerable change. Competition for market share has intensified, and differentiation and market segmentation have become increasingly pronounced (Johnson 1987b, Ford 1991, Foord et al. 1992). Market segmentation, involving specialist retailing for particular client groups, has altered the ways in which retailers operate. Retailers have both expanded their existing markets and developed new ones. At the same time, the development of single stores selling a wide range of goods, described as scrambled merchandising, has been equally apparent. The expansion of market share associated with retail concentration and the expectation of the investment markets has also resulted in a significant trend towards the internationalization of the activities of the largest retail firms (Treadgold 1990). This is already evident in most parts of the developed world.

Recent retail change has taken place in the context of a free-market ideology (Hague 1991). Thatcherism, dating from the 1979 Conservative election victory, sought to relax the constraints imposed by the planning system in a way that inevitably reduced its scope (Thornley 1990).

Following the recession of the late 1970s and early 1980s, the economic recovery, combined with the loosening of planning constraints and administrative fragmentation (Jacobs 1992), contributed to the dramatic restructuring of geographical space. The 1980s has in fact been described as a period of more rapid change in the economic geography of the UK than any other this century (Champion & Townsend 1990). Retailing played a key part in this - compared with the bleaker conditions of the early 1990s (Fyson 1991), the 1980s was a golden age for retailers, when the spatial changes occurring strengthened trends dating back to the 1960s. Perhaps more than any other aspect of retail change, the spatial transformation reflects the complex array of contemporary social, economic, technological and policy trends. For these reasons it merits special attention.

The spatial transformation of retail structure

Until as recently as the mid-1960s the retail system of British cities had a traditional hierarchical structure. Characteristically, this focused on the central business district, which provided the largest concentration of the most specialized high-order goods and services for the surrounding urban region. This was complemented by a relatively small number of local town centres or district centres offering primarily a strong convenience-goods shopping function and a secondary range of comparison goods for specific sectors of the city. At the lowest level, a large number of small neighbourhood centres and corner shops provided a narrow range of convenience goods for the immediate residential population. From the planning perspective this system was considered capable of providing retail services efficiently for the vast majority of the urban population; it was viewed as a system worth maintaining.

Since the mid-1960s, however, suburbanization and counter-urbanization, increasing affluence, rising levels of car ownership and increased female participation in formal employment have allowed and in some cases necessitated the development of new patterns of shopping behaviour. Combined with the demands of consumers for an increasingly specialized and sophisticated range of goods and services, and associated changes in the economic organization of the retail industry, these changes have instigated a transformation in the character of the urban retail system.

Retailers were quick to recognize the commercial advantages of satisfying changing consumer demand by developing large, easily accessible retail outlets on cheap out-of-centre sites. Thus, despite a strong presumption by most local planning authorities against new forms of retail developments that deviated from the traditional hierarchy (Davies 1984), steady commercial pressure has resulted in the emergence of a number of innovatory forms of retailing in most major cities. Schiller (1986) identifies a number of "waves" of retail innovation that have added new elements to the functional complexity of the retail geography of British cities. These new elements can be categorized broadly into five different types.

- The superstores (2,325–4,650 m^2, 25,000–50,000 sq ft) were first developed in the mid-1960s and focus primarily on grocery retailing. Development was most active between 1977 and 1990; by the early 1990s, 600 such stores transacted 20 per cent of the British grocery trade (DOE 1992a). Hypermarkets, which were even larger (greater than 4,650 m^2 or 50,000 sq ft), were also initiated in the late 1960s, with trade almost equally divided between groceries and non-food products. However, few such developments occurred because of the cautious attitude of central and local government planners.
- Retail warehouses (greater than 930 m^2 or 10,000 sq ft) were developed from the late 1970s, initially selling bulky DIY products from premises deemed too large and functionally unsuitable for locations in the traditional shopping centres. Retailers of electrical goods, furniture and carpets quickly recognized the commercial advantages of cheap, accessible, out-of-centre locations, and many such developments exploited the "bulky product argument" as an exceptional reason for a decentralized site. Subsequently, the array of goods sold in retail warehouses has expanded dramatically and includes car accessories, toys, footwear and clothing.
- Retail warehouse parks consist of concentrations of three or more retail warehouses located along a main road or grouped on industrial estates, either as unplanned agglomerations or, occasionally, forming planned concentrations sharing access roads and parking facilities (Bernard Thorpe & Partners 1985). Where such developments incorporate a superstore, a more balanced shopping facility results, and the term retail park has been used for this variant (Wade 1985). The scale of both types of park varies considerably and some of the largest, such as the Swansea Enterprise Zone retail park (about 37,200 m^2, 400,000 sq ft), are similar in scale to a small regional shopping centre (Bromley & Thomas 1989). More than 2,000 retail warehouses were trading in 1992 and approximately 250 retail warehouse

parks were recognized. Together, they are estimated to transact 14 per cent of retail turnover in the UK (DOE 1992a).

- Subregional shopping centres (18,600–37,200 m^2, 200,000–400,000 sq ft) typically incorporate a superstore, at least one large non-food retailer and a number of smaller units in an integrated development. The Culverhouse Cross centre on the western edge of Cardiff, which includes a Tesco superstore and a Marks & Spencer store, is a recently completed example.
- Regional shopping centres (greater than 37,200 m^2 or 400,000 sq ft) are planned as fully integrated, environmentally controlled covered malls. They incorporate department stores, the full range of multiple stores typically represented in a traditional central business district and, in most recent proposals, a significant element of leisure activities. By early 1993 four such centres were trading: the MetroCentre (Gateshead), Merry Hill (Dudley), Meadowhall (Sheffield) and Lakeside (Thurrock). However, a latent demand for many more is suggested by the 57 proposals for centres of up to 139,500 m^2 (1.5 million sq ft) made during the period 1982–91 (DOE 1992a). The commercial impetus for the decentralization of the wide range of quality comparison-goods retailing that characterizes particularly the latter two categories of developments is considered by Schiller (1986) to date from the proposal by Marks & Spencer in May 1984 to open out-of-town stores.

Throughout the period of change, despite the apparently strong official presumption against out-of-centre retail development, a large number and wide range of new types of retail facilities have emerged. This constitutes a major paradox in British retail planning. The early divergences from the hierarchical ideal tend to reflect the circumvention of local planning regulations in a variety of ways by retailers. Subsequently, for most of the 1970s, the intermittent relaxation of control by central government, followed by an increasingly "free-market" orientation throughout the 1980s and early 1990s explains the paradox. Together, the resulting changes in the retail geography of British cities have been considered to constitute a retail revolution (Dawson 1983, Thomas 1989, Gayler 1989b) that is fundamental to the contemporary issues of retail change covered in this book.

The contents of the book

The book is divided into four principal parts, beginning with an economic emphasis, developing via a spatial and planning orientation and progressing towards a number of social issues. The full spectrum of spatial scales, ranging from the international through the national and regional to the urban, are covered and the final chapter highlights the principal contemporary issues that emerge from the analyses of the processes of retail change.

The changing organization of retailing

Part 1 considers the changing organization and internationalization of retailing. Retail concentration and the associated internationalization of retail companies and their formats has emerged as one of the most potent forces of change shaping the nature and character of the sector in western Europe and North America. Growing number of retailers are trading across national boundaries and shops are increasingly selling imported products.

Chapter 2 provides an overview of three key aspects of internationalization. First, the international sourcing of goods, including business consumables and services, is examined by looking at the factors affecting buyer decisions to source internationally, the facilitating rôle of technology and the creation of associations and alliances for sourcing. Secondly, the extent of international retail operations is considered, with an investigation of the reasons for moves to international operations, the mechanisms, and the character of managerial decision-making involved. Thirdly, it examines the international transfer of management expertise, including new ideas and technologies.

Chapter 3 looks specifically at the process of concentration of capital in the British grocery sector in the 1980s, and examines the significance of the benign regulatory environment. An analysis of the national implications of the process focuses on new store development programmes, retailer–supplier relationships, labour productivity and net profit margins. A second major theme is the relationship between the process of retail concentration and internationalization of business organization. The analysis is illustrated by a case study of Sainsbury's expansion into New England on the eastern US seaboard. This is followed by an examination of the entry of European limited-line discounters into the UK, using the examples of the German Aldi and Danish Netto companies.

Both chapters in Part 1 also consider current knowledge about the process of retail internationalization, and the requirements of a future research agenda.

Transformation and the urban region

Part 2 covers retail transformation in the urban region and focuses on two specific aspects: the planned shopping centre and the city centre. New shopping centre development has emerged as a major force in the retail revolution. Chapter 4 shows how these planned retail phenomena have been established at scales ranging from the large regional centre to the smaller retail parks and speciality centres, usually in out-of-town locations but which include city-centre developments. Across Europe as a whole, the significant contrasts in the occurrence and scales of shopping-centre developments raise issues about the effects of differing economic, social and institutional contexts. The relative concentration of retail distribution is influential, while planning constraints have been more effective in curbing developments in some countries than in others. Retailers, developers and investors are participating in extensive cross-border shopping-centre development. Disparities between countries are being reduced, while the co-ordination of planning and environmental policies at the European Community level will have inevitable consequences. These issues are reviewed in the context of contemporary western European experience.

Chapter 5 examines the changes occurring in the city centre. Losses in accessibility because of congestion associated with increases in car ownership, the competition of new out-of-centre facilities of ever-increasing scale and the changing social composition of the inner suburbs have been potent forces for morphological and functional transformation. Central-area redevelopment and festival shopping are now becoming increasingly familiar features of traditional central business districts, usually capitalizing on a combination of internal markets, heritage tourism and a broad city-regional centrality for the development of administrative functions. In detail, however, the process of change varies considerably between countries and cities. This chapter focuses on the differences in the experiences of Cardiff in South Wales and Charlotte, North Carolina, to elucidate the major forces at work. Attention is focused on the influence of central government directives, municipal planning, the local business community and the investment activities of financial institutions to explain the variations. Finally, the significance of these findings for a wider understanding of the transformation of the city centre is developed.

Planning implications of retail change

Part 3 considers the planning implications of retail change. Chapter 6, while recognizing the variations in the recent experiences of the city centre discussed in the previous section of the book, focuses on the evidence of city-centre decline. It examines the changes occurring in the central business districts (CBDs) of UK cities, setting these changes in the framework of the earlier inner-city declines in North America that subsequently spread to Europe. The traditional retail functions of the CBD are examined in the context of the growing problems of central-area traffic congestion, the continued trend towards the relaxation of planning controls on out-of-centre retail developments, and the apparent lack of confidence among potential financiers in the prospects of central business district regeneration. The specific case of the Sheffield CBD is considered, where the decline in retail provision is related to the opening of the large Meadowhall regional shopping centre within a few miles of the city centre. The general problems and future prospects of the British CBD are discussed as major issues of contemporary retail change. A similar theme is pursued in Chapter 7, which looks at the impact of out-of-centre retailing. The proliferation of the new forms in recent years has resulted in a far wider range of shopping opportunities and a longer trading week. This has created a new competitive environment and has precipitated shifts in consumer preference and new patterns of shopping behaviour. The behavioural division between those who use cars and those who do not has been fundamental in this respect. Adverse effects have been suffered by those city centres and other large traditional shopping centres that lack adequate traffic circulation and parking facilities. This chapter reviews the changing patterns of shopping behaviour suggested by studies of the impact of superstores, retail parks and the new regional shopping centres. Evidence is drawn from a number of large-scale household and city-centre surveys that focus on the impact of retail change on the patterns of consumer allegiance and their commercial and planning implications.

Chapter 8 examines planning and shopper security. It focuses on the planning aspects of security problems faced by shoppers and other users of city centres. In recent years fears for shopper safety have reached crisis proportions in some Western cities. The perceived danger of robbery or physical assault is having a seriously adverse effect on some shopping centres. Hardest hit have tended to be the central business districts and other traditional shopping centres in the inner city and inner suburbs. Areas off the main pedestrian routes, multi-storey car parks, pedestrian subways, metro stations and public transport vehicles have all been associated with the rising tide of violence linked with the changing social geography of the inner city. Similar sites, even in the newer planned shopping centres are giving growing cause for concern. The commercial implications are clear. Those centres or parts of centres perceived to offer a safe shopping environment are likely to be most attractive to shoppers. This chapter assesses the significance of this issue for changing patterns of shopping behaviour, both between and within shopping centres, and its implications for the planning and redevelopment of shopping centres. Particular planning strategies are needed to make cities safer, and the chapter reviews some examples of successful policies and practice from recent experience in the UK and the USA.

Social issues of retail change

Part 4 concentrates on the social issues of retail change. Chapter 9 focuses on the disadvantaged consumer. The manner in which the retail revolution, in particular the proliferation of the superstore, has benefited the majority of the population by offering greater choice, comfort and cheapness in shopping is examined. However, superstores are very unevenly distributed and are not equally accessible to all consumers because of substantial disparities in income and mobility. The existence of "disadvantaged consumers" is now recognized as a major social issue. The disadvantaged include low-income households, women, ethnic minorities, the elderly and the disabled, all of whom share the common characteristic of low mobility. The chapter outlines the variety of planning policies adopted to combat the problem of the disadvantaged consumer, policies which include better public transport provision, the protection of local shops, and improved design and facility provision in shopping centres. Finally, the chapter examines the introduction of electronic home shopping and assesses the advantages this technology offers to the disadvantaged consumer.

Chapter 10 focuses on working conditions and the trading week. Changing work practices throughout the retail sector have resulted in the polarization of the workforce between a small number of "managers" and a large number of semi-skilled workers, many of whom are female or part-time. Also, extended shop opening hours and Sunday trading have resulted in increased unsociable working hours. This chapter reviews these changes and examines their implications for the workforce. Particular attention is directed towards the difficulties women experience because of their family obligations and the problems encountered as a result of their dual rôles as carers and employees. Evidence is drawn from European experience and the results of recent attitudinal research are analyzed. The social implications for the family and the manpower policy implications for retailers are discussed.

Chapter 11 investigates the rôle of shopping as leisure. Leisure has always been a component part of the shopping trip; yet, in the past this has not been catered for explicitly by retailers or planners. With increasingly discerning and affluent shoppers, the leisure aspect of the shopping trip is likely to exert a stronger influence on consumer preferences than has hitherto been the case. It seems likely that the centres which do not respond to this challenge will loose ground to centres planned or redeveloped to meet this demand. This chapter explores the characteristics of leisure shopping and draws on a new survey of shopping centre managers to assess the growing importance of leisure provision in planned retail developments. The key components of a leisure shopping environment are identified and their wider implications for the changing characteristics of retail facilities are examined.

Chapter 12 examines the "Greening" of shopping. The spectacular rise in environmental awareness and the emergence of the Green movement has implications for those researching the behaviour of both retailers and consumers. This chapter reviews the many motives for the growth in Green retailing and Green consumption in the UK, some of which have little to do with environmental concerns. Green retailing may be based on a newly acquired ecological altruism or on the fact that retailers and producers have been quick to exploit the lucrative market opportunities for alleged environmentally sound products. The chapter identifies the principal motives for retailers' Green strategies. Looking at the consumer, the chapter reports on a questionnaire survey designed to explore consumers' awareness of environmental problems, the extent to which they have modified their purchasing behaviour as a result of Green awareness, and their motivations. Green consumers vary between those concerned to reduce global environmental degradation and those who adopt healthy eating habits, or merely the veneer of a Green life style. The Greening of shopping is seen as an issue of widespread concern and contradiction.

The book ends with an overview of the contemporary issues of retail change (Ch. 13). An attempt is made to synthesize the main findings of the book and to highlight the principal conclusions.

CHAPTER TWO
The internationalization of retailing

John A. Dawson

The international activities of retailers are apparent in three ways. First, there is the long-established international sourcing of products and services both for resale and for use by the retailer. Secondly, there is the operation of shops, or other comparable retail outlets, in more than one country. Historically, these two types of activities were combined in the activities of pedlars and merchants who travelled from town to town and fair to fair across Europe, buying and selling products as they travelled. In modern times such activities still exist on a small scale, but represent a tiny element of modern retail activity. The two activities of buying and selling now have separate international dimensions. Thirdly, there is the transfer of management expertise from one domestic retail system to another.

Retailing involves creating an assortment of products from a variety of sources and offering them to consumers. Skill and efficiency in doing this is one of the main sources of one retailer's competitive advantage over another. The assortment may be very wide, as in the full-line department store with its several hundred thousand different items or may be quite narrow, as with the 300 or so lines carried in a discount grocery food store. In few cases, however, will all the items in the range, whether it is wide or narrow, have been sourced from domestic suppliers. International buying activities have been for many centuries important to retailers in the creation of their assortments. In most countries in recent years imported products in shops have increased in number. One of the functions of the retailer is to provide the consumer with access to products from outside the residence region, or country, of the consumer. In general, the volume of international trade in consumer goods has risen in recent decades; import penetration in the UK has increased (Morgan 1988) and so international sourcing of products has similarly grown. With the growth in volume of activity so the ways of organizing international sourcing have changed. The first part of this chapter considers some of these changes.

The second main aspect of the internationalization of retailers is in the operation of shops, or other types of retailing, in more than one country. Although some stores were operating internationally by the late 19th century (Mathias 1967) and there were various small international ventures during the first half of the 20th century, a substantial movement towards the international operation of stores did not occur until the late 1960s. The amount and variety of forms of this dimension of internationalization has increased in recent decades, not least because of the increase in size of retail firms and attempts to reduce trading barriers between countries.

Retailing is a very open activity in which it is possible for a retailer to visit and see the operations of other retailers. Relatively little, other than brand names, is subject to copyright and patent. There are well known and well documented cases of retailers having visited other countries and returned to incorporate borrowed ideas in their own company. Examples include a visit to North America by Simon Marks in 1924 (Rees 1969), H. Gordon Selfridge's copying of ideas from Marshall Field's in Chicago, the modelling of the David Jones' store in Australia on Whiteley's of London (Ferry 1960), attendance by European retailers at NCR (National Cash Registers) seminars in Dayton, Ohio, in 1957 and 1960 (Thil 1966), British Productivity Council visits to the USA in the 1950s (BPC 1953), and the visit to 7-Eleven convenience stores in the USA by Masatoshi Ito of Ito Yokado. With the increase in large firms in retailing and the adoption of strategic planning by these firms, so the international search for new ideas and technologies plays a larger rôle in the general internationalization of retail activity. This feature of the international transfer of management expertise is considered in the third part of this chapter.

Although three aspects of internationalization are highlighted in this chapter there are other dimensions to international activity by retailers. International investment by financial institutions occurs such that large retailers are part owned by financial groups, such as pension funds and assurance companies, in many countries. Movements of these financial holdings impinge only indirectly on day-to-day store operations but are still part of the internationalization of retailing. Another area of international activity is cross-border shopping, which sustains some small retail systems such as that of Andorra. These cross-border shopping flows are usually based on price differentials but can be reinforced by relative product availability, with, for example, the international flow since the late 1980s of Hungarian shoppers to large shopping centres (e.g. Shopping Centre Sud at Vosendorf) and hypermarkets in Austria. Of lesser importance are international flows of senior managers, but because of the importance of the chief executive within retail companies with centralized management, such movements are highly significant. Key executive moves into the UK from USA and Australia during the 1980s are particularly noteworthy.

Sourcing in an international context

Buyer decisions to source internationally

Buying is one of the cornerstones of retail activity. With more than 70 per cent of the value of British retail sales being accounted for by the purchase cost of goods for resale, the requirement to purchase effectively underpins all retail activity. There are few surveys of the value of imports in retail sales but one of Safeway by Shaw et al. (1992) indicated that 39 per cent of products by retail value were from non-domestic suppliers. The purchase price of products is one of several criteria used by retail buyers to evaluate whether or not to purchase a product. Price differentials between UK and foreign sources are one factor associated with buying from a non-UK supplier. Other criteria affecting buying decisions are variety, quality and availability. Each has international dimensions.

While price is not necessarily the most important variable, it remains important for many products. In some product areas the price differentials resulting from different production costs are considerable, resulting in substantial non-UK sourcing by British retailers. Toy production, for example, has moved from low-cost production in Hong Kong and Taiwan to even lower cost producers in China, Thailand and the Philippines. Mass-market small electrical and electronic products, mass-market clothing and basic hardware items, often produced to a retailer's specifications, are produced in the low-labour-cost countries of the Far East. International sourcing of these low-price products has been widespread for many years.

A buyer may source internationally in order to introduce variety into the product range being offered to the consumer. One of the aims of the buyer is to create a product range that is different from competitors' but meets consumer needs. The buyer is therefore constantly searching for new products and this includes the products of foreign suppliers, which may bring variety by their foreign origins. The introduction of foreign cheeses into UK supermarkets in a substantial way during the 1960s was partly a response to buyers' searches for variety, as well as consumer demand for more "exotic" items. More recently, foreign beers have had a similar rôle.

Strong foreign brands may also generate the need to source internationally. It is difficult to imagine, for example, operating a successful DIY store without Sandvik handtools or a toyshop without Lego. Although in both cases the manufacturer operates a sales and distribution division in the UK, nonetheless, Sandvik and Lego brands would be considered by most British retailers as products of non-UK origin, since they are manufactured outside the UK. While in these cases the issue is not specifically a search for variety, the activity of the buyer in building a range of products results in the need to source internationally.

The third criterion affecting buyer decisions is quality. The buyer is expected to match product quality to the market position of the retailer. The consumer may also carry perceptions of quality associated with the country of origin of a product (Gaedeke 1973, Bilkey & Nes 1982, Wang & Lamb 1983, Morganosky & Lazarde 1987) and the buyer must be aware of these. To obtain products of appropriate quality may necessitate widespread and international searches for suppliers, and may involve working closely with foreign suppliers in the development of production methods. In the textile and garment sectors, for example, large retail firms often provide the design of garment for the supplier and also specify the fibre, fabric and dye to be used; in addition the retailer will monitor quality in the production unit (UN 1985). Such procedures typify product sourcing for items carrying the retailer's own brand. Access to major manufacturer brands may be important, with some of these probably being of non-domestic origin. Shaw et al. (1992) point out that most manufacturer brand products imported by Safeway are premium brand items.

The fourth major criterion for the buyer is product availability. For the larger retailers it may be necessary to buy from non-domestic producers in order to obtain required quantities and delivery requirements. The growth in size, if not in number, of larger retailers means that internationalization of sourcing is increasingly important. Quota restrictions in the international trade system introduce barriers to large-scale expansion of imports from a single source and sourcing from several countries may be necessary. For example, Silbertson (1990) shows the impact of the Multi-Fibre Arrangement on import sourcing of textiles and clothing into the UK. The widespread reductions in tariffs and quotas under GATT and through the formation of the European Community's Single Market, and the creation of bilateral trade agreements, have all helped to reduce the barriers to international sourcing by retail buyers and encouraged greater product variety in stores (NCC 1991). For larger retailers, use of brokers and agents may be effective in organizing the flow of products and ensuring that delivery requirements are met.

There is an interesting potential reverse issue concerning the internationalization of sourcing, highlighted by Drucker's (1958) account of Sears, Roebuck's activity in Mexico. Rather than a domestic retailer becoming more international by extending its sourcing to new countries, Drucker suggests a process in which a retailer establishes a branch store in a foreign country and then, by encouraging local sourcing for the branch store, stimulates local producers to make goods of specifications of higher quality and to operate more efficiently. This type of internationalization seems less appropriate for the 1990s than it did for the 1950s; now retailers have the possibility of becoming involved in the production processes of foreign-based manufacturers without having to establish stores in the country.

The organization of international sourcing

The search by retail buyers for potential suppliers in other countries may be time-consuming. Buying-offices may be established in other countries. Most large Japanese retailers have buying-offices in London, Paris or Frankfurt; Otto Versand, a German-based mail-order retailer has a network of 22 buying offices, 10 of which are in Asia. Staff in the buying-office will search the market and also act as their parent retailer's representative in dealings with suppliers. The buyer may also visit international trade shows in order to search the market. In other cases a wholesaler or broker may be used. Specialist wholesalers develop detailed knowledge about producers in many countries and can act as a mechanism for retailers to obtain appropriate quality items or to introduce new products at acceptable prices. In the toy sector in the UK, for example, the wholesale firm of Robineau has substantial contacts with toy producers in the Far East, and the use of this wholesaler by smaller independent retailers allows them to source products of appropriate quality and price from low-cost producers.

The buyer in reality is responding to changes in consumer demand and behaviour that indicate increased demands for a greater variety of product. In order to meet this demand retail buyers have to change parts of their ranges more frequently and to introduce new products. Wider searches for suppliers are necessary and foreign products are introduced to meet the new demands of consumers. Buying policy becomes, in consequence, more international.

The facilitating rôle of technology

The increase in international sourcing has been facilitated by developments in information and communications technologies (Burt & Dawson 1991). One of the effects of these technologies has been to reduce the friction of space between retailer and supplier. It has placed distant suppliers in faster contact with the buyers. The use of electronic data interchange (EDI) in which computers communicate directly with each other, and the establishment of international EDI standards, notably EDIFACT, has speeded international order placement and allowed closer matching of retailer orders to production capacities. Developments that link computer-aided design facilities in the retailer, through the retailer buying systems, to computer-aided manufacturing in the supplier are creating a "seamless" supply chain in which distance has little impact until the final products are transported. The US department store retailer J. C. Penney, for example, has introduced technologies that enable buyers and designers in its Dallas head office to link directly with agents and suppliers in East Asia. Design, sample production and product adjustment procedures are communicated electronically so that several days are removed from the design-production-supply cycle for clothing products. Internationalization of supply sources using such technologies depends on the wider availability of the technology, but the technology is not culture-specific and barriers to introduction are typically ones of cost. Such technological developments ease and encourage international sourcing of products by retail buyers.

Buying groups and alliances

A further factor facilitating and encouraging international sourcing by retailers is their willingness to create associations and alliances for sourcing. While alliances may be formed for many reasons (Dawson & Shaw 1992) the importance of the buying function in retailing makes buying activities probably the single most important factor in alliance formation and in their success. These buying alliances are usually of one of three types. A buying group may be created by a group of domestic retailers that wish to combine their buying power and in so doing become large enough to source internationally, organized through the management of the buying group. A typical example is the Toymaster buying group in the UK, one of whose functions is international sourcing on behalf of members. Although present in all European countries, this type of activity is particularly strong in Germany (Tager & Weitzel 1991) and Spain (IRESCO 1984). Growth has also been strong in Italy; Iacovone (1990) points to an increase in number in the non-food sector from 36 in 1972 to 133 in 1987.

Secondly, national alliances of this type may join with others in other countries to enable even more effective sourcing. Several such consortia of national buying groups exist in Europe in a variety of retail sectors. For example, in 1992, Expert operated in the electrical appliances retail sector in 13 European countries, and Ironside in hardware goods trade in 10 countries. Spar covers more than 20,000 shops worldwide but has a particularly strong presence in Europe (Table 2.1). While international sourcing is only one of the activities of these extensive buying and marketing groups it is an important one. The growth of such groups has expanded the international sourcing undertaken by members, who generally are small firms.

A third type of buying alliance has also expanded international retail sourcing. This comprises one retail member from each of several countries. Usually the members are large retail companies. Associated Marketing Services, for example, links ten large grocery retailers, each from a different European country (Clarke-Hill & Robinson 1992). This type of buying group has been present for many decades. The Nordisk Andelsforbund (NAF) alliance of consumer co-operatives, for example, was established in 1918. The late 1980s was a period of rapid creation of these buying alliances in European food retailing. The situation in mid-1992 is shown in Table 2.2. Their formation was a response to such factors as:

- opportunities, including joint buying arrangements, provided by the move to a Single European Market;
- a requirement to source more effectively from non-domestic suppliers;
- a reaction to manufacturers' attempts, notably through mergers, to

increase their size and hence improve their bargaining position with retailers – this is part of the continuing process of creating countervailing power in the European food industry (Segal-Horn & McGee 1989; Dawson & Shaw 1990, Traill 1989);

- attempts to improve the development, including sourcing, of retailer brand products, and;
- probably a copy-cat syndrome after the first few had been established.

Membership of several of the alliances has changed during their brief periods of existence and there is considerable difference in the extent to which the different alliances are successful in co-ordinating purchasing activities.

Table 2.1 Number of stores in Spar 1991.

Country	< 250m^2	250–400m^2	400–1,000m^2	1,000–2,500m^2	> 2,500m^2
Argentina	135	58	5	1	
Austria	964	236	250	37	26
Belgium	123	61	42		
Denmark	285	15	10		
Finland	764	350	245	37	34
France	470	80			
Germany	4,854	855	525	221	69
Greece		3	26	13	
Korea	63	15	8		
Ireland	236	20	4		
Italy	1,653	250	245	65	11
Japan	1,576	64	65	1	
Netherland	76	62	148	4	
Slovenia			3		
South Africa	5	85	272	94	2
Spain	1,254	500	198	10	2
Switzerland	5	11	4	1	
UK	2,336	131			
Zimbabwe	43	5	13		

Source: Spar.

Table 2.2 International alliances and buying groups in the European grocery sector 1992.

Group	Date of origin	Membership	Other linked groups
Associated Marketing Services (AMS)	1989	Ahold (NL) Argyll (UK) Casino (F) Allkauf (D) Hagen Gruppen (N) ICA (S) Kesko (SF) Rinascente (I) Mercadona (E) Migros (CH)	ERA
Buying International Gedelfi/Spar (BIGS)	1990 Gedelfi 1913 Spar 1932	Spar (D, UK, DK, A, I) Dagab (S) Unigro (NL) Tuko (SF) Unil (N) Gedelfi (D) Dansk Supermarked (DK)	Intergroup trading
Cooperation Européenne de Marketing (CEM)	1989	Conad (I) Crai (I) Edeka (D,DK) UDA (E) Booker C&C (UK)	
Deuro	1989	Metro (D/CH) Makro (NL) Carrefour (F)	
Distributeurs Français (Difra)	1968	Casino (F) Casal (F) Catteau (F) Louis Delhaize (B) Francap (F) PG (F) Rallye (F) SCA (F)	AMS ERA
Eurogroup	1988	GIB (B) Vendex (NL) Rewe (D) Coop (CH) Paridoc (F)	

Group	Date of origin	Membership	Other linked groups
Eurocoop	1957	17 consumer co-operative unions in Europe	
European Marketing Distribution (EMD)	1989	Markant (D, NL) Selex (I, E) ZEV (A) Nisa Today's (UK) Superkob (DK) Uniarme (P)	NAF
European Retail Alliance (ERA)	1989	Ahold (NL) Argyll (UK) Casino (F)	AMS
Independent Distributors Association (IDA)	1989	Centra (E) Distributa (D) Huyghebaert (B) Karsten (NL) Superquin (IRL) Tiuron (E)	
Interbuy (IBI)	1988	Intermass (CH) Asdeka (Asko) (CH) Bergendahl (S)	
Intercoop	1971	Konsum (A) CRS, CWS (UK) FDB (DK) EKA (SF) plus 15 smaller national co-ops	NAF
Intergroup Trading (IGT)	1973	National Spar head offices in 10 countries	
Nordisk Andelsforbund (NAF)	1918	Coop (I) CWS (UK) EKA (SF) FDB (DK) KF (S) Konsum (A) NKL (N) SOK (SF) SIS (Iceland)	Inter-coop

Source: trade press.

Models of international sourcing

Much of the previous work on international sourcing relates to the manufacturing sector, where buying decisions are significantly different from those in retailing. Differences exist in:

- the number of items sourced - usually much larger with retailers;
- the number of suppliers used - generally larger with retailers;
- the executive power of the buyer - generally higher with retailers;
- the criteria used to evaluate both a product and a supplier.

There are, however, similarities in the common need to ensure regular supply of products from sources, irrespective of location (Fagan 1991, Monczka & Trent 1991, Min & Galle 1991).

It is possible to postulate a simple two-dimensional stage model of sourcing by retailers. One dimension comprises four stages. The first stage involves domestic sourcing and the use of wholesalers for products of foreign origin. The second stage involves some foreign sourcing using agents working on behalf of the retailer. The third stage involves the establishment of foreign buying-offices to act as facilitators for head-office-based buyers. The fourth stage involves the creation of a worldwide network of buying-offices that supply information, check product quality, organize transportation, and offer a full integrated sourcing service.

The second dimension comprises three stages. There is a first stage in which retailers operate on their own and have an independent sourcing activity. A second stage involves working with others in a domestic alliance to pool purchasing power and to seek economies of scale and of information search by sharing some buying activities with other retailers. A third stage involves establishing an alliance involving retailers from several countries.

Retailers may be operating at different points in this two-dimensional structure. Generally, they move from lower stages to higher stages, but the two dimensions are conceptually independent of each other. While such a model can be postulated from the limited published materials available and anecdotal evidence from retailers and suppliers, there is a need for research to test such a model, and then to refine or replace it in the light of more rigorous empirical evidence.

Sourcing goods not-for-resale and business services

While material on the international sourcing of products for resale is slight, there has been even less research on international sourcing by retailers of capital goods, business consumables and business services. Difficulties of definition occur in attempts to address this area. For example, early suppliers of equipment (e.g. NCR tills) and current suppliers of information technology (IBM, ICL, etc.) operate internationally and it is difficult to know what constitutes a foreign supplier because discussions, negotiations, payment, after-sales support, etc. may well be undertaken by the local (national) office. Similar issues arise in business services with, for example, the large international accountancy services operating through local offices.

Despite these definitional difficulties, it can be surmised that retailers are undertaking more international searches than formerly for supplies of a wide range of capital goods, such as materials-handling equipment and checkout equipment, and for consumable office materials and business services. Although we can suggest a general increase in international sourcing, there are examples of reduced activity: for example, it is interesting to note the retreat from the use of tropical hardwoods in shop fitting in much of Europe and their replacement by other materials, in some cases domestically produced. The criteria used for evaluation of suppliers may not be the same as those used in the evaluation of sources of supply of goods for resale. It is also possible that alliances and information networks are important contributors to the search and evaluation processes. Without research in this area, however, it is difficult to do more than surmise on what may be the situation.

International retail operations

Definitions

Retailers, if they are to be successful, have to respond to the culture of their customers. Any move towards multi-store operations has to acknowledge that culture varies through space. This is particularly apparent when retailers choose to operate in diverse cultures, whether within one country or across national borders (Martenson 1987, 1988). International retail operations may be defined as the operation, by a single firm, of shops, or other forms of retail distribution, in more than one country. Many of the issues associated with operating in more than one country are the same as those that affect operations within culturally different, and sometimes autonomously governed, regions of the same country. For example, within Europe the issues and different requirements associated with operating stores in both the Autonomous Regions of Andalucia and Catalonia in Spain may be less onerous than those associated with operating stores across the national frontier between Belgium and Holland. The cultural differences of consumers, employees and business practices are less in the international context than in the national one in this example. More examples of this type may occur in Europe as the Single Market mechanisms become effective and currency convergence begins to be a wider reality but at the same time regional consciousness in social issues and cultural identities becomes more overt (Burt 1989, Dawson 1991). Nonetheless, it is useful to isolate international retail operations as a particular case of retail activity because national borders still have meaning for the political, governmental and judicial environment in which retailers operate.

The international operations of retailers have been subject to substantial study since the early reviews of European (Knee 1966) and American (Yoshino 1966) activity and the more substantial studies by Carson (1967) and Hollander (1970). Experimental moves, not all successful, by some larger European retailers were considered briefly by Dawson (1978) and more substantially by Waldman (1978), while attempts by American retailers to help in the modernization of retailing in developing countries were considered by Goldman (1974a,b). Although there was some activity by retailers in exploring new national markets before the mid-1980s, the project to remove the physical, technical and fiscal barriers to trade in Europe by 1 January 1993 encouraged retailers to broaden their horizons and in turn stimulated research to look beyond trans-Atlantic (Kacker 1985) movements, and to consider trans-border moves within Europe. It is with these more recent international moves that the remainder of this section is concerned.

Reasons for moves to international operations

Dunning (1981) in a general consideration of direct foreign investment (DFI), irrespective of sector, suggests three factors that together are of importance in establishing whether a firm develops direct investment in international operations (i.e. establishes shops of its own). If present individually, they generate different forms of indirect internationalization (Dunning & Norman 1985).

First are ownership-specific advantages: a firm has an innovatory product, process or business method that gives it competitive advantage in the market. Examples from retailing (with which Dunning does not deal explicitly) are retail brand products (Body Shop, Laura Ashley) or an individually refined sales method (Benetton, Aldi, McDonalds). These advantages may be obtained by licensing or franchising without DFI.

Secondly, there are location-specific advantages: a potential host country has particular cost advantages or market opportunities not present in the home country. For retailing, differentials in the cost of land and labour or differentials in market growth and existing market structure illustrate this factor. The differentials between France and Spain in both respects are specific examples (Treadgold 1990a). Again most of these benefits can be obtained through indirect investment methods.

The third factor comprises the internalization advantages in which the organization of the firm or some environmental factor results in ownership and locational advantages only being released through DFI. Examples of the constraints might be legal impediments to franchising, an absence of firms suitable to be licensed, unfamiliarity with the business concept in the host country, the appearance of acquisition opportunities, etc. Sears, Roebuck's direct investments in Latin America were apparently influenced by the lack of indigenous managerial expertise, while Carrefour's hypermarket development in the USA may well have been perceived in the same way. Ahold's stores in the USA resulted from suitable acquisition opportunities being present (Kacker 1990, Hamill & Crosbie 1990).

Dunning's (1981) three factors provide a framework for considering retail internationalization but they also serve to highlight the considerable differences between DFI decisions in retailing and manufacturing sectors and even between the organization and management of firms in the two sectors (Pellegrini 1992). The balance between centralized and decentralized decision-making, the relative importance of organizational and establishment scale economies, the degree of spatial dispersion in the multi-establishment enterprise, the relative size of establishment to the size of the firm, the relative exit costs if decisions are reversed, the speed with which an income stream can be generated after an investment decision is made, different cashflow characteristics, the relative value of stock and hence importance of sourcing; all these items, and others, serve to differentiate the manufacturing firm and the retail firm not least in respect of the internationalization process. These differences call into question suggestions (Whitehead 1992) for the direct application to retailing of the stages (Aharoni 1966, Wilkins 1974) and network (Hakansson 1982) approaches, and theories of internationalization (Buckley 1983) and DFI developed for manufacturing firms (Welch & Luostarinen 1988, Leontiades 1985).

Reasons provided from the several empirical and conceptual studies (Alexander 1990, Treadgold 1991, Treadgold & Davies 1988, Salmon & Tordjman 1989, Tordjman & Dionisio 1991, Burt 1991, Laulajainen 1987, 1991a) for the establishment of foreign retail operations are:

- perceived current or imminent market saturation in the home country; this factor has been suggested as particularly important when the home country has a relatively small market, such as Holland or Belgium, and so is relevant to the international moves of Ahold, GIB and Vendex;
- presence of an unexploited market or growth market in the host country; this factor has been suggested as important in moves during the 1980s into Spain particularly by French retailers and moves into Eastern Europe after 1989;
- higher profits (net margin or return on investment) obtained in the host country because of differences in competitive and/or cost structures (Treadgold 1990a);
- presence of a consumer segment that has not been targeted in the host country but for which a store format has been developed in the home country; this has been suggested as relevant to retailers such as Toys 'R' Us, The Gap, Aldi, Ikea, Disney Shop and Body Shop;
- the spreading of risk across several, possibly unrelated, markets – the diversified holdings of Metro, Tengelmann, and Vendex and the moves by food retailers into DIY retailing are suggested as examples;
- use of surplus capital or the gaining of access to new capital sources at lower cost than in the home country – some moves by Japanese retailers in the second half of the 1980s suggest this factor;
- entrepreneurial vision; Marks & Spencer's moves into Canada appeared to owe much to the enterprise of then top management, the determination of Katsu Wada to internationalize Yaohan and moves by Ratners into USA or by C&A into several countries might be similarly categorized;
- an opportunity to get access to new management ideas or technology that will then be transferred to the home country – this may be important in areas such as use of information technology, having been suggested as one reason for Sainsbury's move to USA and Edvardsson, Edvinsson & Nystrom (1992) suggest it can also affect concepts of customer service in the home country operations;

- consolidation of buying power;
- reaction to the internationalization of manufacturers - this has been suggested as relevant to some of the moves within Europe of food retailers and in the DIY sector;
- encouragement, possibly inducement, by major manufacturer-suppliers wishing to enter new markets in which they currently have limited presence - suggestions have been made that major toy manufacturers have been supportive of moves by Toys 'R' Us into Europe and Japan, and manufacturers of designer-brand products may encourage international expansion by retailers such as Louis Vuitton, Gucci and Dunhill;
- limits placed on domestic growth in the home country by public policy controlling new store development or limiting further growth in market share of a firm; it is suggested that the Loi Royer in France, which was created to limit hypermarket development, encouraged hypermarket operators to look for development opportunities in other countries;
- the desire to follow existing customers abroad; the establishment of small branches of major Japanese department stores in major European capital cities allows loyal customers to continue using the store when on holiday or working in the host country.

This list covers reasons for international moves, not necessarily for successful moves. Success or failure in international expeditions by retailers can be ascribed to many more factors than are listed here.

Dimensions to international operations

The two key dimensions for a typology of international retail operations are the mechanism by which internationalization is achieved and where in the overall organization decision-making resides for the retail business in the host country.

The mechanisms to achieve internationalization of operations may be summarized as:

- internal expansion, in which a company opens individual shops using in-company resources;
- merger or takeover with the acquisition of control over a firm in the host country;
- franchise-type agreements in which the franchisee in the host country use the ideas of the franchiser based in the home country (Eroglu 1992);

Table 2.3 Advantages and disadvantages of alternative mechanisms to establish international operations.

Mechanisms	Advantages	Disadvantages
Internal expansion	Can be undertaken by any size of firm. Experimental openings are possible with modest risk and often modest cost. Ability to adapt operation with each subsequent opening Exit is easy (at least in early stages). Allows rapid prototyping.	Takes a long time to establish a substantial presence. May be seen by top management as a minor diversion. Requirement to undertake full locational assessment. More difficult if host market is distant from the home market.
Merger or takeover	Substantial market presence quickly achieved. Management already in place. Cash flow is immediate. Possibility of technology transfer to home firm. May be used as a way to obtain locations quickly for conversion to the chosen format.	Difficult to exit if mistake is made. Evaluation of takeover target is difficult and takes time. Suitable firms may not be available. Substantial top management commitment necessary.
Franchise-type agreements	Rapid expansion of presence possible. Low cost to franchisor. Marginal markets can be addressed. Local management may be used. Wide range of forms of agreement available. Use locally competitive marketing policy. Way of overcoming entry barriers.	Possibly complex legal requirements. Necessary to recruit suitable franchisees. Difficult to control foreign franchisees. May become locked into an unsatisfactory relationship.
Joint venture	Possible to link with firm already in market. Help available in climbing learning curve. Possible to move later to either exit or make full entry into the market.	Necessary to share benefits. Difficulties in finding a suitable partner.
Non-controlling interest	Find out about market with minimal risk. Allows those who know the market to manage the operation.	Passive position. Investment made over which little influence.

- joint ventures, which may take a variety of forms for the joint operation of retailing, including in-store concessions, between a firm in the host country and one in the home country;
- non-controlling interest in a firm in the host country being taken by a firm in the home country.

Some of the advantages and disadvantages of the different mechanisms are summarized in Table 2.3. In addition to these mechanisms is the purchase at auction of previously state-owned stores. This method is one favoured in Eastern Europe. For example, in Hungary in 1992 there was an auction of 150 shops formerly the property of state chain Kozert; the German company Tengelmann purchased 24 and the Belgium food retailer Louis Delhaize 10. This type of activity may be considered as a special case of the takeover mechanism.

"The internationalization process varies from company to company, but even within one company there can be great differences" (Edvardsson, Edvinsson & Nystrom 1992: 81). A retailer may use more than one method as it develops the scale and variety of its international operations. Marks & Spencer, for example, acquired existing companies in Canada and USA, has developed through internal expansion in France and Belgium, undertook initially a joint venture in Spain, has used franchise-type arrangements in Greece and initially in Hong Kong, entered Hungary with a joint venture before moving to a franchise type arrangement, and had a non-controlling interest in the company that sold St Michael products for several years in Japan. The international expansion of Aldi (Table 2.4) illustrates a more typical combined pattern of expansion, first by internal expansion to nearby and culturally similar countries, then with a major acquisition in a non-European market and, finally, expansion across a broad front involving small acquisitions and single store openings.

The second key dimension to classifying international operations is the extent to which managerial decision-making is centralized at a head office in the home country or is delegated to host country operations. This aspect of management is related to whether the retail operations are virtually the same from country to country (for example, McDonalds, Toys 'R' Us, Aldi, Virgin), or whether the shops reflect the regional and national society in which they operate (for example, Dixons' operation of Silo in the USA, Carrefour's individual hypermarket operations in Spain, Argentina, Brazil, Portugal, Taiwan and USA). Salmon & Tordjman (1989) make this distinction, terming the retailer "global" if there is great similarity in operation from country to country and "multi-national" if there are explicit national differences.

There exists in reality a continuum between the extremes of one global identity and a locally tailored identity but any of the five mechanisms of establishment shown in Table 2.3 may be applied at the global or the multinational sections of this continuum.

Table 2.4 International expansion of Aldi.

Country	Year of first opening	Number of stores, end-1990	Sales (billions DM)	Market share (%)
Germany	1948[a]	2,100	21.5	12.0
Austria	1960	130	1.8	8.5
Netherlands	1975	261	1.3	5.0
USA	1975	225		
Belgium	1976	250	1.8	9.0
Denmark	1977	102	0.6	3.0
France	1988	15		
UK	1989	4[b]		
Spain	Aldi logo used by IFA-Español			
Italy	15 sites being considered			
Poland	Entry anticipated before end of 1992			

Notes: a. First store in former East Germany opened in September 1990; b. By July 1992 there were 43 stores operating.

There are two additional variables that are usually important in classifying retail operations. These are the market position of the retail offer; for example, whether a discount store, premium brand store, mass mail order, etc; and the format of retailing, for example, whether hypermarket, brand concession in a store, convenience store, life-style boutique, mail order, etc. It might be hypothesized that these two variables will affect both the mechanism of international establishment and the position on the global–multinational axis. Perhaps a strong premium brand might lead to a global approach through internal expansion or tightly controlled franchising, possibly exemplified by Louis Vuitton (Laulajainen 1992), Body Shop or Dunhill (Gapps 1987). Alternatively, the large floorspace mass merchandise hypermarket might lead to a multinational approach with joint-ventures and possibly takeover as a favoured way to become established, as exemplified by Promodés. However, such hypotheses can be countered by exceptions to the suggested relationships and, as pointed out above, a single firm may use several approaches to gaining an international presence. There is still a need for substantial research to relate market position and format type to the different forms of international presence (Robinson & Clarke-Hill 1990). The relationships, if they exist at all, are far from clear.

The extent and directions of international operations

Comprehensive information on the extent of the international operation of retailers is not available, but data collected by Corporate Intelligence Group (1990, 1992a) and a database for food retailing in Europe compiled by Burt (1991) provide an indication of recent trends, at least for European retailers.

The data from Corporate Intelligence Group are concentrated on international activities of retailers headquartered in EC countries and of retailers from outside the EC that have established an activity in the EC. Partial information is provided on the non-European international activity of these two groups of firms. The data include activities made from before 1970 until 1991. The database is not complete and excludes many activities made within this twenty-year period that proved unsuccessful. Nonetheless, despite problems with completeness, the data allow interesting conclusions to be drawn:

- A total of 2,057 activities were recorded and ascribed to 459 firms; of these, 1,321 activities were within EC countries and of these 1,090 were undertaken by firms headquartered in the EC.
- There was a notable increase during the second half of the 1980s in the number of international actions by the 459 retailers. This trend continued into the 1990s, with 450 actions recorded after 1990 compared with 371 between 1985 and 1989. Difficulties of tracing the early moves may account in part for these figures but this is only a partial explanation.
- Retailers from France, Germany and UK account for almost 70 per cent of the EC-based firms involved in international actions. These firms account for almost three-quarters of the international actions recorded within the EC.
- Internationalization is more common among non-food retailers than food retailers. Of the 1,321 activities recorded with the EC countries, 869 were in non-foods and 286 in foods. The relatively low level of

international activity of food retailers was also noted by Waldman (1978).

The difference between food and non-food sectors is partly, but not wholly, accounted for by the larger numbers of non-food retailers. Another reason may be that food, more than fashion, has a strong local cultural aspect to consumer choice, which limits opportunities for scale economies in buying to be achieved through international operations. Jefferys (1973), also noting the difference, points to the priority of food retailers, particularly in the 1960s, being the development and exploitation of the new methods and techniques in food retailing associated with supermarkets; and this gave them little time to consider international expansion. Reasons for the relative success of non-food retailers in internationalizing in Europe are listed by Tordjman et al. (in press) as:

- many of the non-food formats are small and require limited capital and managerial cost for their establishment;
- entry and exit are easier with small formats;
- formats often target small consumer segments, which limits expansion in domestic markets;
- single-brand stores give a unique competitive advantage to the format (Williams 1992) which is suitable for international transfer;
- non-food formats are more suited to franchising;
- there are strong economies of replication that can be transferred to the host country;
- the cachet of a foreign retailer is easier to stimulate in non-food than in food retailing.

Burt (1991) confirms several of these trends for food retailing, indicating a similar increase in activity from the late 1980s and the proportional dominance of retailers based in France, Germany and the UK. He also indicates that while British retailers have favoured moves to other English-speaking countries, notably Ireland and the USA (Burt & Dawson 1989), French retailers have favoured Spain and the USA, and German retailers have concentrated on Austria and the USA. A quarter of the 230 actions studied, irrespective of home country within Europe, involved moves to the USA. Wrigley (1989) and Hallsworth (1990a) consider some of the attractions of the USA from a European viewpoint, while Siegle & Handy (1981) view the early moves from a US perspective. The attraction of cultural closeness, at least for the initial international moves of a retailer, are also indicated as important for Swedish retailers (Martenson 1981, Laulajainen 1991b). Burt suggests differences in preferred mechanisms; French food retailers tend to favour joint ventures, minority interests and more co-operative approaches; while British firms have favoured acquisition. Within the food sector most retailers (64 per cent of the 67 companies considered) have been involved in only one or two international actions. The decision to open stores in another country is not taken lightly (Jackson 1973), and a gradualist approach to international expansion is usually taken.

The internationalization of management ideas

The third main route for retail internationalization is through the flows of know-how and management expertise. This has increased substantially in recent decades as awareness has expanded of the activities of retailers around the world. Innovation is a key requisite for the successful retailer and the diffusion of innovation is therefore an important process in the retail sector. The importance of innovation and new ideas encourages wide and international searches to be undertaken.

The transferability of retail concepts

The extent to which a retail concept can be transferred from one culture to another has perplexed researchers, development economists and management consultants for many decades (Savitt 1990). There were many attempts, from the 1950s onwards, to transfer American ideas of supermarkets to the developing economies but few resulted in the expected objectives being met (Goldman 1981). In some cultures the supermarket became used by higher-income groups rather than as an agency of mass distribution to lower- and middle-income groups as was the objective. In other cultures severe resistance was met, not from consumers, but from small retailers, who saw supermarkets as a threat to their own existence. Elsewhere, the absence of physical, social and economic infrastructure created barriers to the transfer of unadapted retail methods. In other cases resistance came from sources concerned about commercial imperialism. Typical of the difficulties in the transfer of ideas are those that were encountered by the International Basic Economy Corporation (IBEC) established in the USA in 1947 to encourage US businesses to transfer management skills to other countries, mainly in Latin America. The organization was involved, through different initiatives, in the introduction of supermarkets to Argentina, Brazil, El Salvador, Peru and Venezuela (Broehl 1968), but by 1976 all had been closed down. Nonetheless, in most cases a form of supermarket has evolved from these, sometimes forced, introductions and the concepts of supermarket retailing have become firmly established in many Third World countries.

Transferability of retail concepts on an international scale in Europe or across the Atlantic poses different, but nonetheless substantial, problems. There are many examples of retailers failing to appreciate the difference in cultures between superficially similar retail environments. The experience of Early Learning Centre in attempting to establish its British concept in North America is instructive. The differences in consumer attitudes to purchasing educational products for pre-school children together with differences in the real-estate market were instrumental in the failure in the USA of a well defined successful concept relevant to the UK. Some retailers fail to appreciate the nature of the competition in the host country. For example, Aldi's entry into the UK and Carrefour's entry into the USA appear to have been resisted by existing firms, which have sought to heighten the entry barriers for these foreign firms, despite the fact that both Aldi and Carrefour have well defined and successful store formats. In other cases institutional and political impediments are introduced to raise entry barriers for specific innovations and, although such barriers may affect domestic retailers as much as foreign entrants, if the innovation is foreign then such activity limits the extent of internationalization of the concept. For example, Japanese attempts to limit the creation of large specialist non-food stores, while affecting Japanese retailers, served to limit the internationalization of innovatory concepts such as those developed by Toys 'R' Us and Laura Ashley.

While the retail concept may be technically open to internationalization, the behaviour of the players in the retail marketplace may inhibit acceptance or change the concept such that it becomes acceptable. There is still, in such cases, an internationalization process at work, but one in which the concept is only one of several aspects affecting the spread of the idea. In such a process it becomes very difficult to establish which characteristics of the concept affect its international transferability. Recent attempts to transfer retail concepts from western to eastern Europe illustrate this dilemma, in that while a technology gap (Kacker 1988) may be bridged relatively easily, the non-retail aspects of the acceptance of innovation limit the transferability of the retail concept.

Types of expertise transferred

The direct transfer of ideas from retailer to retailer can take many forms. The interactions are substantial and extend well beyond the transfer of operational concepts. The types of expertise transferred include:

- store formats;
- design concepts - examples include innovations in display systems and internal layouts;
- management tools and techniques, for example, promotional ideas, productivity measurement methods such as direct product profitability, staff incentive schemes, locational assessment models;
- retail technologies, for example, the diffusion of use of electronic point of sale systems, adoption of management information systems;
- customer services, for example, the spread of frequent shopper programmes, uses of direct customer communications.

In most of these cases, however, there is little intrinsic difference between a retailer copying ideas from a foreign retailer or from another domestic retailer. The Benetton display systems, for example, were copied by retailers both in Italy and in other countries; the customer service provision in Nordstrom has been copied by other retailers in the USA as well as by visiting Europeans. There are few cases where an idea has specific cultural values attached to it and whose transfer involves a process of internationalization of management. Cultural values may be attached to retail products such as Scandinavian, Italian or Spanish designed furniture, but it is unusual for such values to be attached to managerial concepts. The transfer of ideas is likely to increase as firms get larger and formalize their market research and market development activities and it seems likely that the international transfer of ideas will similarly increase as communication techniques reduce the friction of space.

Mechanisms for the international flow of ideas

Channels for the flow and transfer of ideas and the mechanisms to encourage transfer can be formal networks and initiatives or may be more informal, entrepreneurial and individualistic.

A number of mechanisms are influential in international transfer of retail concepts, among which governmental and quasi-governmental initiatives are of primary importance. The single-market initiative by the governments of European Community member states has, for example, had an encouraging influence on the international flows of retail ideas in the late 1980s (EC 1991, Filser 1990), not least by simply raising awareness of the differences in retailing within Europe. Similarly, in Japan the MITI (Ministry of International Trade and Industry) sponsored review and prognosis documents of the retail trade have highlighted the opportunities for Japanese retailers arising from the study of foreign companies. Likewise, the Distributive Trades EDC (DTEDC 1973) attempted to raise awareness of retailing in other European countries in the wake of British entry into the European Community.

Also of importance have been the activities of the trade associations and chambers of trade. A common function of retail trade associations is the provision of a forum for the exchange of information (Knee 1968). Members of these types of groups have opportunities to exchange information and to participate in formally organized visits to study retail operations in other countries. The long history and voluntary participation in these events testifies to their value as a medium for the exchange of ideas. The internal communication within firms having international operations through joint ventures and from participation in international alliances has had comparable effects.

Advisors, consultants and the providers of services also serve to facilitate the flow of ideas. Management consultants on projects and through their publications (for example, Management Horizons' monthly *Retail Europe* and Arthur Andersen's house journal *International Trends in Retailing*) also use the experiences of foreign retailers to provide information and services to retailers in many countries. The international network of offices of such firms serves both to collect and disseminate information. Other mechanisms of this type are IT companies which, being themselves international in operation and as part of their exercise of relationship building with clients, encourage retailers to look at what is going on in other countries.

Finally, individuals are perhaps the most potent mechanism for the international diffusion of ideas in retailing. The movement of entrepreneurs and of managers, whether in establishing new businesses in different countries (Lord et al. 1988), or simply visiting other retailers to glean ideas, inevitably contribute to the international transfer of ideas.

Conclusions

The increase in international retail activity is inexorable. Having been set in motion, the internationalization process seems to have quickened its pace. International sourcing is widespread and commonplace; international operations, while still only undertaken by a small minority of all retail firms, are increasingly common activities within the large firms; the barriers to the free international flow of ideas appear to be lessening and the flows increase in volume and variety year by year. This chapter has explored these three themes.

A substantial amount of research already exists on the internationalization process in retailing, but it is still not clearly understood. The process, for many reasons, is substantially different from the internationalization process in manufacturing firms. While some concepts may be borrowed from the literature on industrial internationalization, they are unlikely to be directly applicable to the retail sector. The structures of the sectors are different, the processes of evolution differ, and the behavioursof the various participants in the process are different. The internationalization of operations is at a relatively early stage compared with other aspects of internationalization and also by comparison with the manufacturing sector.

Within the retail sector a key issue in the internationalization process is the need for the adaption of management practices in response to the cultural character of the host country. For some retailers this adaptation has proved to be particularly difficult, even in moves to apparently culturally similar environments. A second key issue is the important rôle played by individual entrepreneurs able to take an international perspective. A third conclusion to be drawn from the research is that retailers are often unaware of the impact upon, and value for, the firm of the process of internationalization. These issues are likely to figure high on the agenda of future research into the processes of internationalization in the retail sector.

CHAPTER THREE

Retail concentration and the internationalization of British grocery retailing

Neil Wrigley

During the 1980s, Britain experienced a massive and sustained concentration of capital within grocery retailing. The trend towards concentration had been present for the previous twenty years, but the 1980s were characterized, most notably, by the emergence of a small group of retail corporations whose turnover, employment levels, profitability and sheer market and political power came to rival the largest industrial corporations in any sector of the UK economy. Between 1982 and 1990, the market share of the top five grocery retailers increased from under 25 per cent, to 61 per cent of national sales, on one estimate, and a "super league" of just three firms began to separate out in terms of growth, profitability and annual capital investment. During years which have been described as the "golden age" of British grocery retailing, the immense oligopsonistic buying power wielded by the retail corporations came to condition all aspects of retailer–supplier relations, and created new corporatist relationships between the retailers and the regulatory state.

Central to the transformation of corporate power during these years was the spatial switching of retail capital. This took place at three distinct scales.

The intra-urban scale of switching, from traditional inner-city shopping areas to edge-of-city sites, created within less than a decade a new urban landscape. "Cathedrals of consumption", often erected on greenfield development sites whose land values were forced up by planning restrictions and between-firm competition towards £2 million per acre by the end of the decade, became the engines of corporate growth. The inner-city High Streets, the locus of profit extraction during the shift to self-service grocery retailing in the 1960s, increasingly became the focus of disinvestment by the major corporations, neatly exemplifying Harvey's (1985: 25) characterization of the "perpetual struggle in which capital builds a physical landscape appropriate to its own condition at one particular moment in time, only to have to destroy it . . . at a subsequent point in time".

The inter-regional scale of switching witnessed an inter-meshing of the areas of expansion of the major grocery corporations as they moved out from their traditional core areas. By the end of the decade with Sainsbury in the North East and moving into Scotland, Asda in London and the South East, and Tesco in Yorkshire, an increasingly uniform spatial offering characterized food retailing in the UK, reflecting the highly concentrated "big capital" nature of the industry. Both the intra-urban and inter-regional scales of switching of retail capital are well documented in the literatures of geography and planning (Schiller 1986, Davies & Sparks 1989, Jones 1981, 1991, Guy 1988b, Howard & Davies 1988, Sparks 1990, Thorpe 1991).

The third scale of switching – the international scale – however, attracted far less attention, although early discussions by geographers are noteworthy (Treadgold & Davies 1988, Wrigley 1987, 1989, Hallsworth 1990a, Lord et al. 1988, Treadgold 1990b). In many respects this was surprising, as the internationalization of British grocery retailing encapsulated very precisely both the successes and the potential fragility of a massively restructured domestic industry. The purpose of this chapter, therefore, is to explore and attempt to understand the dimensions and implications of that internationalization. Both outward movement of retail capital from the UK, and inward movement of retail capital into the UK, will be considered, and it will be seen that internationalization at the upper and lower ends of the UK food retailing must be viewed as part of a single interrelated process. Outward movement of British retail capital will be illustrated in terms of the transatlantic movement of British food retailing to the USA, whereas inward movement will be illustrated via an analysis of the entry into the UK market of limited-line discount retailers from mainland Europe.

It will be seen that internationalization is a key feature in any analysis of the future trajectory of corporate growth in British food retailing, and that internationalization has, and will continue to have, profound consequences for the competitive structure of the industry. In effect, the UK is now exposed to pressures from which it was largely immune during the 1980s; a decade of intense restructuring in the UK retail industry, characterized by the rise and transformation of corporate power within a benign regulatory environment.

The "golden age" of British grocery retailing

To understand how the internationalization of British grocery retailing came about, it is first necessary to appreciate some of the key features of the retailer-dominated UK food system which emerged during the 1980s, for it was those features which drove that internationalization. Four features will be picked out. In practice these are intimately interconnected, but for purposes of discussion it is useful to separate them.

The concentration of capital and the rôle of the new store development programmes

Although measures of market share are notoriously confused and confusing in this field, the rise of the major grocery corporations in the 1980s can be represented to an approximate degree by Table 3.1. In the early 1980s the top three firms - Sainsbury, Tesco and Asda - held a combined market share of around 20 per cent of national grocery sales. They had recently been joined by two other firms, the Argyll Group and the Dee Corporation (later Gateway), which were growing rapidly by takeover and merger. Between 1982 and 1988 the dominance of this "big five" group increased considerably. Their combined market share reached approximately 43 per cent by 1984 and 58 per cent by the end of 1988. By 1988, Sainsbury, the largest food retailer, had surged up the rankings of British companies by turnover, from 51st position a decade earlier, to 14th position. It had increased its workforce from 32,000 to 88,000 and had sustained net profit rises and shareholder-dividend increases of more than 20 per cent per annum throughout the period.

The late 1980s saw a continuation of this remarkable progress. By 1990, on one estimate, the top five grocery retailers held 61 per cent of national sales, although other estimates, calculated on different definitions of the total national food market, place the figure closer to 50 per cent. However, what is indisputable is that a "super league" of just three firms - Sainsbury, Tesco and Argyll (the operators of Safeway) - had begun to separate out in terms of growth, profitability and annual capital investment. The latter was particularly important, for the critical arena of competition between the major grocery retailers had become the new store development process. By the late 1980s the major grocery corporations were generating no less than two thirds of their annual increases in sales from the new stores they had opened in the previous twelve months, and those new stores were operating at higher net-profit margins. The new store-expansion programmes thus became vital to the maintenance of the substantial annual increases in turnover and profit which the capital markets had come to expect. Given the rationing of sites for large-scale development by planning constraints, competition for the most attractive development sites became intense. The late 1980s were therefore an era of "store wars" in which strategic capital investment and an ability to ground that capital became the engine of corporate growth and ever-increasing concentration.

Table 3.1 Top five British grocery corporations: estimated shares of total grocery sales.

	1982[a]	1984[b]	1988/89[c]	1990[d]
	%	%	%	%
J Sainsbury plc	9.5	11.6	14.5[f]	16.3
Tesco plc	8.7	11.9	14.8	15.7
Argyll Group plc	(3.8)[e]	5.1	9.7	11.2
Asda plc)	4.6	7.2	7.9	10.4[g]
Dee Corporation/Gateway plc	n.a.	7.3	11.4	7.8[h]

a *Source:* AGB figures quoted in Davies et al. (1985: 9).

b *Source:* Verdict Market Research.

c *Source: Retail Business Quarterly Trade Reviews* **12**, December 1989: 12. Figures relate to year to March 1989.

d *Source:* Verdict Market Research (*Supermarketing*, 18 January 1991).

e Crude estimate: Argyll acquired Allied Suppliers in 1982. Allied's market share on acquisition was approximately 3%.

f Does not include grocery sales of Savacentre, the Sainsbury subsidiary. If included, rankings of Tesco and Sainsbury are reversed.

g Prior to major financial and organizational problems of Asda 1991/92 (for details see Wrigley 1991, 1992b).

h Following buyout by Isosceles and disposal of assets to Asda and Kwik Save.

Retailer–supplier relations and the return on capital in food retailing and manufacturing

In the words of analysts at County NatWest Wood Mackenzie, the UK food manufacturing/processing sector "entered the 1980s as a fragmented, inefficient, largely domestically based industry struggling to come to terms with the growing power of the major food retailers" (*The Times*, 9 Feb 1990). Since the 1960s and, in particular, the passage of the Resale Prices Act in 1964, power had moved away from the manufacturers/suppliers and towards a group of food retailers committed to building market share by passing on both the volume discounts which they obtained from the food manufacturers, and the lower operating costs from larger-scale outlets and faster stock turnover, to their customers. Indeed, by the end of the 1970s there was already concern amongst the UK regulatory authorities that the shift in power from food manufacturers to retailers may have gone too far and that the rapidly expanding food retailers might be receiving non-cost-related discounts from the food manufacturers/processors of a magnitude sufficient to constitute an anti-competitive abuse of buying power (Wrigley 1992a).

The events of the 1980s simply compounded this shift in power. The hugely increased market share of the major food retailers gave them immense oligopsonistic buying power which conditioned all aspects of food manufacturer–retailer relations. In addition, the large food retailers aggressively exploited own-label trading and their ability to control the allocation of selling space in order to squeeze out the nationally advertised brands of the food manufacturers (Davies et al. 1986). In combination, these two factors allowed the large food retailers to demand ever-larger discriminatory discounts, including backdated discounts, from the food manufacturers and to impose ever more stringent conditions of supply.

Stringent conditions of supply imposed on food manufacturers to match the centralized, logistically efficient, stock control and quick-response warehouse-to-store distribution systems developed by the retail corporations in the 1970s and early 1980s (Sparks 1986, McKinnon 1989), allowed the large food retailers to pass back more and more of the costs, including "uncertainty costs", of inventory holding to the manufacturers. Moreover, the large food retailers typically operated on negative working-capital cycles (Norkett 1985). That is to say, they were supplied on credit but turned over their stock well before their suppliers had to be paid. As a result, within the accelerated circuits of capital which came to characterize British food retailing, food manufacturers were obliged to pass to retail capital an ever-greater proportion of the surplus value generated in the act of production. The effect of this was seen in both profitability and relative returns on capital employed in food manufacturing and retailing. As Figure 3.1 demonstrates, the return on capital in food manufacturing declined relative to food retailing in the late 1970s and early 1980s. Moreover, during the 1980s, profit levels of the food manufacturers increased at only half the rate achieved by the major food retailers.

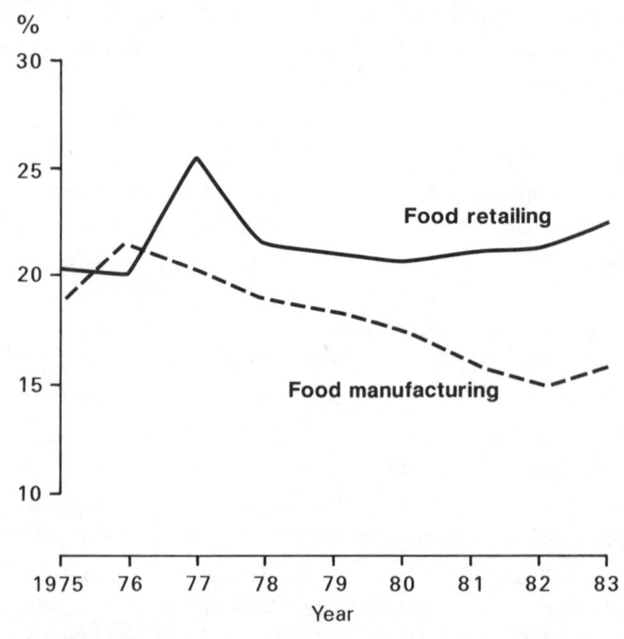

Figure 3.1 Return on capital employed in UK food retailing and food manufacturing, 1975–83. (*Source:* adapted from OFT 1985).

Despite a tendency during the 1980s towards greater concentration amongst British food manufacturing/processing companies, and the emergence of food manufacturers such as Hillsdown Holdings, Albert Fisher, Hazlewood Foods, Cadbury Schweppes, etc. as the largest source of cross-border mergers and acquisitions within the European food industry (and as important players in the North American market) food manufacturers were obliged to learn to live with the realities of retailer dominance. Increasingly they sought stability and communication in their relationships with the major food retailers, relationships in which they could develop effective and profitable strategies for anticipating and responding to retailer demand. Within a "preferred supplier" type of relationship (Crewe & Davenport 1992) between food retailer and manufacturer, retailers increasingly dictated product innovation, development and quality control. Illustrative of this trend was the fact that, amongst food manufacturers, R&D expenditure fell significantly, while the employment of food technologists by the major grocery retailers doubled between 1975 and 1985 (Bowlby & Foord 1989).

Labour productivity and net-profit margins

Labour productivity in British retailing had risen strongly from the 1960s to the early 1980s (Table 3.2). This reflected the transition which had occurred from smaller labour-intensive, to larger capital-intensive, forms of outlet. Also, it reflected an intensification in the use of labour associated with the replacement of a predominantly full-time employed work force by a largely part-time, deskilled, and increasingly feminized work force, matched with ever-increasing sophistication by the large retail corporations to temporal fluctuations in consumer demand. As a consequence of these trends the proportion of female part-time employees in British retailing increased from 18.8 per cent in 1957 to 38.8 per cent in 1983.

Table 3.2 Labour productivity in British retailing 1961–82.

Productivity	1961	1966	1971	1976	1980	1982
Sales per FTE* (£ thousands)	25.78	27.44	31.36	36.42	39.92	40.66

Source: IFS Institute of Fiscal Studies 1984. * Sales are retail sales revalued by retail price index to 1982 prices. FTE is total number employed in retailing converted to full-time equivalents.

Improvements in labour productivity had been particularly marked within the top grocery retailers, and these improvements continued strongly throughout the 1980s, aided by massive capital investment in computer-based information technology (IT) systems. Sainsbury, for example, increased its labour productivity by 14 per cent in real terms between 1978 and 1982 and followed this with a 12 per cent increase between 1983 and 1987. Significantly, however, the period 1983–89 saw Sainsbury's computing power growing at 50 per cent per annum, and the proportion of sales passing through EPOS (electronic point of sale) scanning checkouts rising from 1 per cent to 90 per cent. EFTPOS (electronic funds transfer at point of sale) facilities followed rapidly and were introduced into 74 per cent of stores by 1989, and the mass of capital employed per full-time equivalent employee rose twice as fast as the retail prices index.

The increases in labour productivity achieved by the larger UK grocery corporations, together with the imposition of their oligopsonistic buying power on the food manufacturers were amongst several factors powering a dramatic rise in net-profit margins during the 1980s (Table 3.3). By the late 1980s these margins had risen to levels which were quite unusual in international terms.

Table 3.3 Trends in net profit margins (%) of "big three" grocery retailers' major operations 1985–92* (figures based on turnover exclusive of VAT).

Retailer	1985	1986	1987	1988	1989	1990	1991	1992
Sainsbury	5.25	5.53	6.09	6.54	7.31	7.61	8.32	8.71
Tesco	2.72	3.10	4.10	5.21	5.86	6.18	6.62	7.09
Safeway	3.57	3.89	4.34	4.69	5.19	5.94	6.74	7.49

Source: Henderson Crosthwaite (1992).

* i.e. takes account of only major format/fascia of the company, *not* secondary formats/operations such as Lo-Cost/Presto (Argyll), Savacentre/Homebase/Shaw's (Sainsbury).

Retailer–regulatory state relations

The concentration of capital in British grocery retailing took place in a lenient, pragmatic and benign regulatory environment. The framework of British competition law, and its interpretation by the UK regulatory authorities, the Monopolies and Mergers Commission (MMC) and the Office of Fair Trading (OFT), were conducive to the development of the highly consolidated "big capital" industry which emerged in the 1980s, and to the growth of the oligopsonistic buying power of the grocery retailers. In particular, the findings of two major MMC/OFT investigations of the industry, *Discounts to retailers* (MMC 1981) and *Competition and retailing* (OFT 1985), provided the regulatory conditions for the "golden age" of British grocery retailing (Wrigley 1992a).

During this period, what Flynn & Marsden (1992) have described as a "new corporatist relationship . . . between government and the increasingly concentrated retail sector" developed. This relationship in which the government often enlisted the major grocery retailers as "agents and promoters of public policy" for example, in the politically sensitive area of food safety regulation, was "crucially dependent upon the continued economic dominance of retailers as the new masters of the food system" (Flynn & Marsden 1992: 92).

Internationalization as a response to changing competitive conditions

The spectacular performances of the major grocery retailers in the 1980s brought in their train a certain fragility of position and, as the 1980s came to an end, that fragility became both more obvious and more significant for the internationalization of the industry. Two aspects of this issue are worthy of attention.

Capital markets, enhanced risks and the treadmill

The engine of corporate growth during the 1980s was strategic capital investment in new store-expansion programmes. The replacement of "non-conforming" (often High Street) space by new more capital-intensive space in out-of-centre superstores "ensconced and dominant in their own market(s) . . . enjoy(ing) a degree of protection from all" (Moir 1990: 112) provided one of the pivots on which the increases in labour productivity and net-profit margins rested. However, as the decade advanced, those store-expansion programmes became progressively more expensive in real terms. In the case of Sainsbury, for example, annual capital expenditure increased no less than tenfold during the 1980s to over £500 million per annum by 1989, an increase five times greater than the retail prices index. By 1991/92 it had increased to £800 million per annum. The new store-expansion programme of Tesco rose even more markedly and had reached £1 billion per annum by 1991/92.

Despite net-profit margins which had risen (Table 3.3) to quite unusual levels by the end of the 1980s, even the hugely increased pre-tax profits being achieved by the "big three" grocery retailers were rapidly becoming insufficient to fund these expansionary programmes from retained earnings. Yet there was no simple way of disengaging from a corporate trajectory ("chosen path of accumulation") which was threatening to become a treadmill. The capital markets had, over the decade, come to expect and discount continued substantial annual increases in turnover and profit, yet those annual increases have been shown to be intrinsically dependent upon the new store-expansion programmes. As a result, "failure to meet these exacting annual new store-expansion targets (could) have a marked effect on year-end results, and any slip (was) compounded by failure to meet unrealistically demanding stock market expectations" (Wrigley 1987: 1286).

By the beginning of the 1990s two of the "big five" grocery retailers, Asda and Gateway, had indeed fallen victim to the demands of this exacting treadmill. In the case of Asda, failure to drive forward the critical new store-expansion programme with sufficient pace resulted, in 1989, in an attempt to compensate in one leap via the purchase of 61 superstores from Gateway for £700 million at the time of Gateway's leveraged buyout by the Isoceles consortium. The purchase, aimed at gaining market share in critical regions of the UK, proved ill advised. It overextended the firm and, burdened by debt as the recession of the early 1990s began to bite, Asda's performance crumbled. The capital expenditure programme had to be cut back savagely, profits declined, its share price fell dramatically, its Chief Executive left the company, and by 1992 it was announcing an annual pre-tax loss of £365 million.

As Gateway and Asda faltered and dramatically curtailed investment, a massive gap opened between the "big three" grocery corporations (Sainsbury, Tesco and Argyll) and the second-tier retailers. In 1991, at the bottom of a services-led recession in the UK, the "big three" raised £1.4 billion of new capital in a wave of public rights issues (Wrigley 1991) to fund their accelerating expansion programmes. They were also active raising capital in the bond and sale-and-leaseback markets. Pumping over £2 billion per annum into store-expansion programmes, the "big three" announced plans to increase their floorspace by 8–9 per cent per year during the 1990s, thus doubling their physical capacity by the year 2000. Yet the sheer scale of this investment served merely to raise even more strongly the doubts that had been growing in the late 1980s about the long-term viability of this strategy and the return on capital which would be achieved by the major grocery corporations. The *Financial Times* (19 June 1991: 22) raised the question of "how long the likes of Sainsbury, Tesco and Argyll can continue to reap attractive returns from their enormous investment programmes", and ventured the opinion that:

> "The trouble about food retailers is not the immediate outlook; it is rather the way all possible good news is taken for granted. One of these days, it must surely all go wrong".

Similar doubts were expressed publicly by growing numbers of retail analysts in the City of London, and were given increased substance by the sharp divergence of values which had emerged between the food-retail property market and the general UK property market in which property values had fallen by over 20 per cent (Wrigley 1992b). Indeed, Shiret (1992) suggested that, if the overvaluation of the grocery retailers' property was corrected via depreciation, and a procedure known as the "capitalization of interest" was discontinued (Wrigley 1992b), the pre-tax profits of the "big three" would decline by between 10 and 20 per cent, placing yet more question marks against the accelerating store-expansion programmes.

By the late 1980s and, more starkly, during the deepening recession and property crisis of the early 1990s, it became clear, therefore, that the hugely increased funding requirements of the major grocery corporations were raising significant doubts about the trajectory of corporate growth in British grocery retailing. Clearly, if the major corporations were not to become victims of their past successes in the "golden age" of the 1980s, they needed to look to sustain their past levels of performance in new ways. Diversification, into other sectors of retailing, or into grocery retailing outside the UK, offered obvious routes. In particular, internationalization offered the opportunity to establish bases which could "be used defensively to provide opportunities for capital investment in an era in the mid-1990s . . . when UK-based profits growth at the levels which the stock market has come to expect and discount may become more difficult to sustain" (Wrigley 1989: 288).

High margins, excess profits and price competition

As the profit margins of the major grocery retailers rose to ever-higher levels in the late 1980s, the conditions of competition in the UK grocery market came under closer scrutiny. This came first from overseas retailers, envious of the much higher margins being obtained in British grocery retailing (particularly in comparison to mainland Europe and North America). They sensed the emergence of opportunities for aggressive low-margin, low-capital-intensity, operations. Secondly, scrutiny came from retail analysts, consumer groups, and the press, who were all becoming increasingly alarmed at the growing disparity between UK profit margins in the grocery sector and those elsewhere. In particular, they were suspicious that ever-increasing concentration in UK grocery retailing within a benign regulatory environment had resulted in conditions supportive of the extraction of "excess" (shared monopoly) profits.

In the popular press, articles drawing attention to the high net-profit margins, lack of real competition, and the "scandal of Britain's high food prices" to finance "the building of cathedral-like supermarkets, as store chains compete to . . . increase their control over the retail food market" (*Sunday Times*, 25 August 1991) started to appear. In the trade press the grocery corporations were reminded that they were "turning in the highest net profits in the world" and cautioned to remember the issue of "real price value" (*The Grocer*, 10 August 1991). Such articles coincided with increasing academic concern over shared monopoly and the possible extraction of "excess profits". Moir (1990: 93–7), for example, noted that the years following the period (1979–83) studied in the OFT's *Competition and retailing* investigation were characterized by very different conditions: by marked and sustained increases in both gross and net-profit margins. He also suggested that a pattern of increasing gross-profit margins and increasing net margins sustained over time was strongly indicative of the possibility of "excess" profits.

In response, the major grocery retailers maintained adamantly that the increased margins merely reflected added value and changed commodity mix. Furthermore, they argued that, in comparison with overseas grocery retailing (in particular North America), higher margins were necessary in Britain to repay the higher investment required on expensive British land, and that the true comparison should be on the ROCE (return on capital employed) measure, which was very similar in British grocery retailing to elsewhere. Nevertheless, the exposed margins of the major grocery retailers had become a matter of public concern, and into the debate were drawn Members of Parliament and members of the House of Commons Select Committee on Agriculture (Wrigley 1991). An OFT or MMC referral began to seem a possibility. In addition, there was sensitivity to the fact that a tightening of the benign regulatory environment in which the major grocery retailers had operated might flow from any UK government commitment to reform the Fair Trading Act 1973 and the Competition Act 1980 along the lines of Article 86 of the Treaty of Rome in pursuit of wider EC harmonization objectives.

In many ways the conditions of the early 1990s were ideal, therefore, for the penetration of the UK market by overseas retailers stressing a "value platform" of aggressive pricing and low margins. First, the established "big three" grocery retailers had adopted a corporate strategy of raising entry barriers by hugely increased investment and differential access to capital. With the minor exception of Argyll's Lo-Cost operation, they had totally abandoned the low-margin discount sector and, despite their protestations, the "price wars" of the late 1970s were merely faded memories. The low-margin, low capital intensity, limited-line discount sector of the market which they had abandoned had, in turn, developed its own corporate giant, Kwik Save, the 6th/7th ranking UK grocery retailer. The corporate performance of Kwik Save in the 1980s had been immensely impressive. Pre-tax profit had grown by 19 per cent per annum in the 5 years to 1990, accompanied by the highest return on capital employed (ROCE) in the UK food retailing sector. However, Kwik Save had grown in a relatively protected market niche. Despite its low-margin, low operating-cost image, and a corporate policy of re-investing profit-margin gains into competitive prices, its net-profit margin of 5.9 per cent in 1990 was high by international standards. In addition, it had begun, quite distinctly, to move away from its initial, restricted-range, very low-capital-intensity, format towards a broadened product range, slightly higher cost operation (County NatWest WoodMac 1991a). Moreover, it was clearly untested by serious competition offering a similar "value platform" in the same market niche. That is to say, the possibility existed that its impressive growth was partly an artifact of being the only strong player in a rather protected niche within a general food sector which was, arguably, lacking "real" competition. Furthermore, unlike the high entry barriers presented by the capital-intensive strategies of the "big three", the low capital intensity of the Kwik Save format presented only low barriers, of both entry and exit, to new competition.

Secondly, the enhanced risks of the high capital-investment trajectory of the major corporations had left the two weakest members of the old "big five" in a seriously debilitated state, struggling to service major debts of £678 million in the case of Asda, and over £1 billion in the case of Gateway, in 1992. Thus, the second tier of the industry was potentially highly vulnerable, particularly in urban/industrial regions in which consumers were becoming increasingly price sensitive as the deepening recession of the early 1990s took hold. Asda and Gateway found themselves at a critical juncture without the capital necessary to mount either a challenge to the "big three" or to defend themselves with sufficient vigour against determined low-margin challengers in increasingly "value sensitive" urban/industrial markets.

Thirdly, the bouts of public concern over the excesses of retailer dominance of the UK food system which surfaced in the early 1990s left the incumbent grocery retailers highly exposed over the issue of the benign regulation of competition in the industry and threatened the corporatist nature of retailer–regulatory state relations.

Into this potential market gap rode the European limited-line discounters, notably Aldi and Netto. They were attracted to the UK by the logic of their own long-term corporate expansion trajectories and by static (saturated) home markets. Undoubtedly, they were also attracted by the high ROCE it appeared possible to achieve in UK discount food retailing, and by general "margin envy". High and consistently rising profit margins achieved in Britain suggested to these European food retailers, as much as to the UK press and some UK academics, the extraction of excess profits. Indeed, some retail analysts have suggested that the European limited-line discounters may have succumbed to some sort of "conspiracy theory" as the major reason for rising UK profit margins (Kleinwort Benson 1990), allied to the influence of the retailers' oligopsonistic buying power. However, whatever the reasons for the attempt by the European discounters to penetrate the UK market in this way, they were clearly sensitive to public concern over conditions of competition in the UK grocery sector, and willing to test and exploit the rather exposed relations between the retailers and the regulatory state. Indeed, by 1991, two formal complaints to the OFT had been lodged by Aldi, and the challenge to the sensitive relationship between the incumbent retailers and the regulatory authorities was being used as a market entry tactic.

Two facets of the internationalization of British grocery retailing

Internationalization as a response to the competitive conditions of the "golden age" of British grocery retailing became focused, therefore, upon two distinct segments of the market. At the top end, the highly capital-intensive strategy of the "big three", allied to the quasi-monopolistic nature of UK planning permissions, raised entry barriers to impregnable levels. However, increasing doubts began to centre upon the long-term return on capital from the seemingly ever-accelerating superstore-expansion programmes. The need to consider diversification grew more pressing. In particular, capital investment opportunities in growth areas were required for an era, in the later 1990s, when expansion opportunities in UK food retailing could be expected to decline, capital expenditure requirements would fall, but the huge cash-flows from the multi-billion turnovers of the top corporations would continue unabated and that capital would need to be "grounded".

By contrast, at the bottom (discount) end of the market, low capital intensity and high ROCE were the key to market success. However, this in turn implied low entry barriers to new competition. The "golden age" of British grocery retailing created a UK grocery market with many apparent attractions for foreign retailers, and a combination of conditions in the early 1990s was highly conducive to market penetration from a segment of the industry which had developed particular strengths in mainland European markets.

Two case studies which illustrate these themes will now be considered.

The lure of the USA: Sainsbury's expansion into New England

By 1992, Sainsbury (in the guise of Shaw's) had become a major food retailer in the northeast USA, operating a network of 73 stores, with a further 6 stores under development (Fig. 3.2). Its turnover in 1991/92 had reached $1.81 billion. Despite extremely depressed trading conditions in New England during the recession of the early 1990s, Sainsbury was widely reported as being "very bullish" about the prospects for Shaw's which would "help take on the engine of group profit growth as and when UK development opportunities become more limited" (Smith New Court 1991: 47). To this end, Sainsbury was planning to accelerate the pace of its US expansion.

To understand how this position had been arrived at, it is first necessary to appreciate the lure of the US market to a major UK grocery retailer such as Sainsbury, seeking avenues for diversification and opportunities for capital investment. In simple terms, the fundamental attraction of US food retailing for British retail capital lay in the marked disparities which had emerged by the late 1980s between the British and US industries, most notably in corporate concentration, retailer/supplier power relations, net-profit margins, labour productivity and the relative returns on capital to retailers and manufacturers/suppliers. When allied to financial factors such as the freedom of UK capital from exchange controls, the emergence of the USA as the major global recipient of foreign capital inflows to fund its burgeoning budget deficit, the strength of sterling against the dollar in the late 1980s, the ability to buy control of US companies through an "open" stock market, and certain advantages conferred by pre-1990 UK accounting rules (notably the ability to write off the "goodwill" acquired during a takeover against shareholders' equity within the first year of operation) – these dispar-

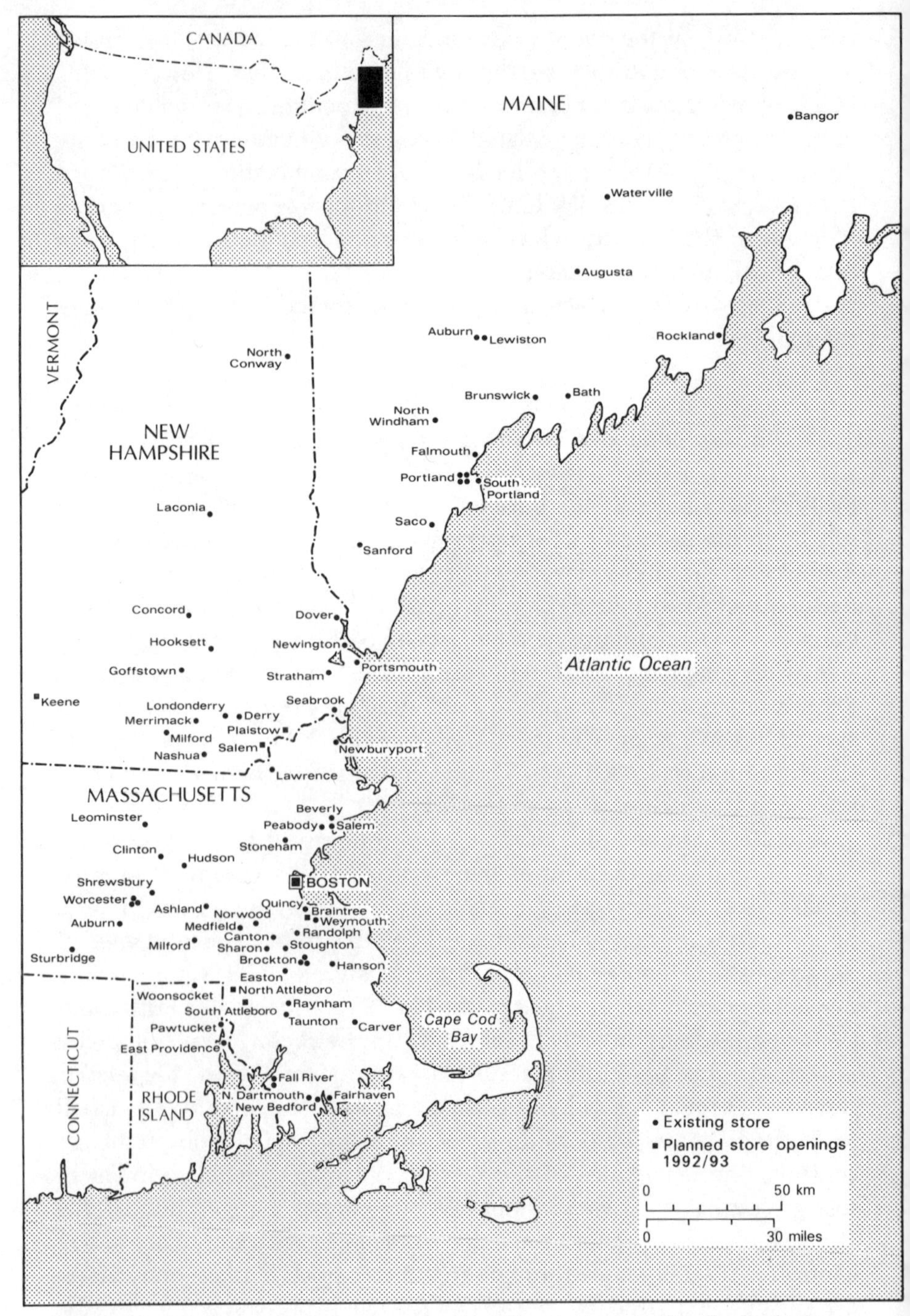

Figure 3.2 Sainsbury in the USA: the network of Shaw's stores in New England. (*Source:* information supplied by J. Sainsbury plc.)

ities offered considerable opportunities to the major UK grocery retailers. In addition, and not to be underestimated, was the attraction of investment and operation in an English-language business environment.

Space in this chapter does not allow a detailed consideration of why significant disparities had emerged between British and US food retailing by the late 1980s. That story is outlined elsewhere (Wrigley 1992a). It is sufficient to note that an industry had emerged in the US which was characterized by: considerably less concentration of capital than in the UK (Table 3.4); a far less significant shift in power from the food manufacturers/suppliers to the food retailers than in Britain; less ability on the part of the retailers to pass back the costs of inventory holding to the manufacturers/suppliers; higher returns on capital employed in food manufacturing than in retailing (a complete reversal of the situation in Britain, shown in Fig. 3.1); flat or declining labour productivity compared to sharply rising UK labour productivity; and much lower net-profit margins (Table 3.4). For a period of almost 50 years, from the 1930s to 1983/84, US grocery retailing had operated in a regulatory environment which was hostile to the development of "big capital" in retailing and to market share being concentrated into the hands of a few large chains operating multi-regionally and enjoying considerable purchasing leverage. An industry had emerged in which the dominant position of food manufacturers/suppliers and their margins was protected, and which became characterized by a geographical structure of regional chains.

Table 3.4 Top five US grocery retailers in the mid-1980s: sales, market share and net profit margins.

	1985 Sales		Net profit margin (%)	
Retailer	$ million	share (%)	1985	1986
Safeway Stores Inc.	19651	7.0	1.2	0.9
The Kroger Co.	15967	5.7	1.1	0.3
American Stores Co.	12119	4.3	1.5	1.1
Lucky Stores Inc	9237	3.3	1.0	0.9
Great Atlantic and Pacific Tea Co.	8049	2.9	2.8	1.2

Source: Wrigley (1992a), Litwak (1987).

Arguably, by the mid 1980s, "inefficiency" was evident in US grocery retailing. In this context, Moir (1990: 93–7) has suggested that a pattern of increasing gross-profit margins but declining or stable net-profit margins, sustained over a period of time is indicative of rising costs relative to sales and the emergence of "inefficiency". Exactly this pattern can be observed in US grocery retailing in the early 1980s. Gross-profit margins were rising (Marion 1986; Fig. 5.3) but net-profit margins were at best stable, hovering around 1 per cent (Table 3.4), or in some years declining. However, the perception among certain financial institutions was that the industry was capable of improving its net-profit margins, and that leveraging, possibly of a majority of the major firms in the industry, would do much to correct the inefficiency and improve performance (Magowan 1989). As a result, and for other reasons relating to consistent cash-flows and readily saleable undervalued assets, in the more permissive regulatory environment which developed after 1984, grocery retailers became prime targets for leveraged buyouts (LBOs). Thus, the industry became enmeshed from 1986 to 1989 in what Clark (1989) has termed the "arbitrage economy" (see also Hallsworth 1991a).

The LBO period of the late 1980s encouraged a certain degree of consolidation within the industry, plus an upward trend in net-profit margins as the LBO firms became more focused, efficient, and drove tougher bargains with unionized labour. However, LBOs also had the effect of imposing massive debt burdens on the firms involved. This deprived them of capital for investment in store development, computer-based IT systems etc., just at a time when the major UK grocery corporations were positioning their long-term competitive strategy on hugely increased investment and differential access to capital.

By 1987, it was clear that the attractions of the US market to British grocery-retailing capital were becoming increasingly apparent. The chairmen of two of the "big three" grocery corporations made well publicized statements to the effect that:

> "Tesco is anxious to expand abroad should the right opportunity come along. Group executives have been looking at the US market with acquisitions in mind for some time." (Ian MacLaurin, Chairman, Tesco plc, quoted in *The Times*, 1 October 1987)

and

> "In the longer term we will consider widening our retail operating base. This may involve investment in food or other retailing in the US where we have been keeping a watching brief over the past three years." (James Gulliver, then Chairman, Argyll Group plc, *Company report* 1987: 7)

Meanwhile Sainsbury, the largest corporation, made a decisive move into US grocery retailing via a $261 million purchase in June 1987 of full control of Shaw's Supermarkets, a US retailer operating 49 stores in New England in which Sainsbury had previously built up a 28.5 per cent holding. Significantly, Sainsbury's purchase was announced as being part of a "long-term strategy of reducing dependence on British food retailing" (Evan Davidson, Treasurer of Sainsbury's quoted in *The Times*, 20 June 1987). The Shaw's buyout was followed immediately (October 1987) by acquisition of Iandoli, a Worcester (Massachusetts) chain of supermarkets, adding a further 10 stores to the network.

During the next five years, Sainsbury developed Shaw's, in what rapidly became very difficult trading conditions in the recession-hit New England economy, using techniques very similar to those which underpinned its organic expansion in Britain. In particular, a capital-intensive strategy was followed, and investment was targeted in the same manner as in the UK into: (a) a vigorous store-expansion programme, (b) the development of a logistically efficient, quick-response distribution system, and (c) the installation of integrated computer-based information systems incorporating stock control, sales-based ordering, distribution, administration, and financial control, etc. In addition, Sainsbury imported into a US food-retailing environment characterized by much lower levels of own-label trading (approximately one-third the levels in the UK) an intensive programme of own-label product development. By 1990, 300 own-label products had been introduced, and by 1992 the range had been expanded to 800 products and was showing significant volume growth. Finally, Sainsbury adopted in Shaw's many of the labour practices that characterized its UK operations, and signed, after considerable opposition, a new deal with the United Food and Commercial Workers Union in 1991. Between 1989 and 1992 the total number of Shaw's employees increased at a much lower rate than the real increase in sales turnover and, significantly, the proportion of part-time, contingent, workers in its labour force rose from 67.8 per cent to 74.8 per cent, with a commensurate 7 per cent fall in full-time employees to just 25.2 per cent.

Table 3.5 provides some of the key statistics on the store-expansion programme. It should be noted that the relatively small increase in total store numbers from 1988 to 1993 disguises a very active programme of replacement, enlargement, remodelling and closure of existing stores. Indeed, the sales area of the chain increased at approximately 9 per cent per annum, identical to the annual increase planned for Sainsbury floorspace in the UK for the 1990s.

Of equal importance to the store-expansion programme was the attention Sainsbury gave to up-grading Shaw's distribution and supply-chain management along UK lines. Particular attention was paid to: the introduction of UK-type central buying methods; third-party contract

Table 3.5 Shaw's store expansion programme 1987/88–1992/93.

Year[a]	1988	1989	1990	1991	1992	1993 (est.)
Sales area (,000 m^2)	148	157	179	196	207	226
(,000 sq ft)	1,592	1,693	1,928	2,107	2,229	2,430
Increase on previous year (%)	–	6	14	9	6	9
Stores	60[b]	61	66	70	73	79
New stores opening in year	–	6	8	6	3	6

Source: Sainsbury annual reports.
[a] Trading year to March of stated year.
[b] Includes acquisition (October 1987) of 10 Iandoli stores.

distribution; the improvement of Shaw's data systems to permit the computerization of direct store delivery receiving, truck dispatching and routeing; and the development of a UK-type chilled supply-chain. These measures involved investment in new distribution centres (a 27,050 m^2 (291,000 sq ft) perishable goods distribution centre was opened in 1991), and provided the opportunity to test the largely untapped potential in the US grocery market for prepared and chilled foods.

By 1992, five years after taking control of Shaw's, Sainsbury had laid the foundations of a quality regional food retailing operation in the US. As a proportion of the total Sainsbury Group turnover and profit, Shaw's was still relatively small. Its turnover in both 1991 and 1992 was about the size of the Waitrose or Wm Morrison food retailing operations in the UK, and it contributed on average only 5 per cent of Sainsbury group profits. However, Sainsbury was clearly in a position to carry Shaw's through a protracted period of market-share building and to invest the capital necessary to build a US retailing base with considerable strategic potential for the late 1990s. Shaw's "longer term prospects are thought to be really very encouraging" (Smith New Court 1991). Moreover, the development of Shaw's has challenged one of the prevailing, and often repeated, myths of the emerging literature on the internationalization of retailing. For example, Burt (1990: 13) has stated that: "the consensus view appears to be that standardized concepts and operating policies do not travel well in the case of food retailing". Yet Sainsbury is clearly building its Shaw's operation in a successful fashion around UK trading, distribution, and supply-chain management techniques. What this demonstrates, is the danger of making over-generalized statements on the spatial switching of retail capital without embedding these statements within a contextual analysis of the dynamics and regulation of corporate restructuring.

The entry of European limited-line discounters: Aldi and Netto's expansion into the British market

The background to the European limited-line discounters' interest in the UK grocery market, and the potential gap which had opened in the market, has been outlined above. Rumour of impending entry became fact in April 1990 when Aldi, the major German discounter, opened its first store in Birmingham. Aldi was quickly followed in December 1990 by the entry of Netto, the discount operation of the Danish retailer Dansk Supermarked. It was also believed that two other German discounters, Norma and Lidl, were about to begin operations in the UK. In the event, Norma and Lidl both shelved any plans for expansion into the UK. However, Aldi and Netto began to build their UK operations and, by 1992, they had developed the store networks shown in Figures 3.3 and 3.4. Together, they accounted for slightly over £200 million, or less than 1 per cent of total UK grocery sales, and operated from only 68 stores. This compares with almost 800 stores and sales of £2.3 billion in the case of Kwik Save, the major UK discounter. However, their initial impact on the UK grocery market, particularly that of Aldi, had been out of all proportion to their actual size.

Stated simply, the fears generated by Aldi's penetration of the UK market among the incumbent retailers lay in Aldi's reputation for aggressively low pricing, based upon extremely low gross margins and severely constrained operating costs. This was allied to Aldi's competitive strength and massive international buying power which flowed from being one of Europe's largest retailers, operating 2,900 stores in mainland Europe, with a global turnover very similar to that of Sainsbury, and turnover of DM22 billion, or approximately 12 per cent of the German market (County NatWest WoodMac 1991b, Smith New Court 1991). In addition, it was known that Aldi was prepared to make no significant profit for many years when entering and building its share of trade in a new market. Moreover, when compared to Britain's high-profile, extensively analyzed, and publicly traded grocery retailers, Aldi was a privately controlled and rather secretive corporation, owned and run by the Albrecht family. With a stated target of developing 200 stores in its UK operations, and fears that the real target might be 500, it was estimated that Aldi might rapidly gain between 2.5 and 5 per cent of the total UK grocery market. Trading at this level, Aldi would be capable of inflicting considerable competitive damage on the existing UK discounters and convenience/neighbourhood grocery store operations such as Kwik Save, Lo-Cost, Co-op, Budgens, Spar/VG, and Circle K,

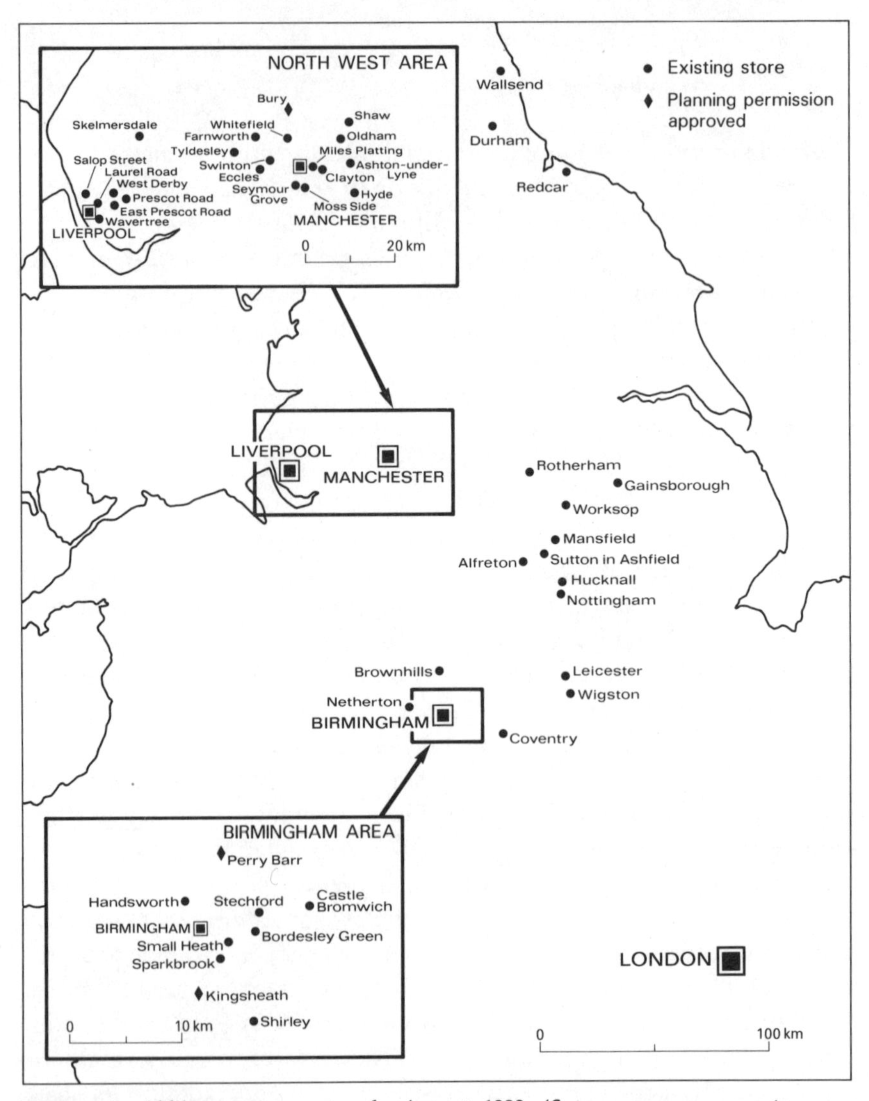

Figure 3.3 Aldi's UK store network, August 1992. (*Source:* company reports, press announcements of store openings, and planning application listings.)

etc.; and possibly also on the vulnerable second-tier grocery retailers, Asda and Gateway. Moreover, faced with a competitor willing and able to trade on gross-profit margins of well under 10 per cent, and net margins of perhaps only 2 per cent, several commentators became extremely concerned that entry of Aldi into the UK market would destabilize the rather comfortable margin structure of the UK industry, and inject price-competition ripple effects throughout the market.

The inflamed sensitivities of the incumbent UK retailers were hardly

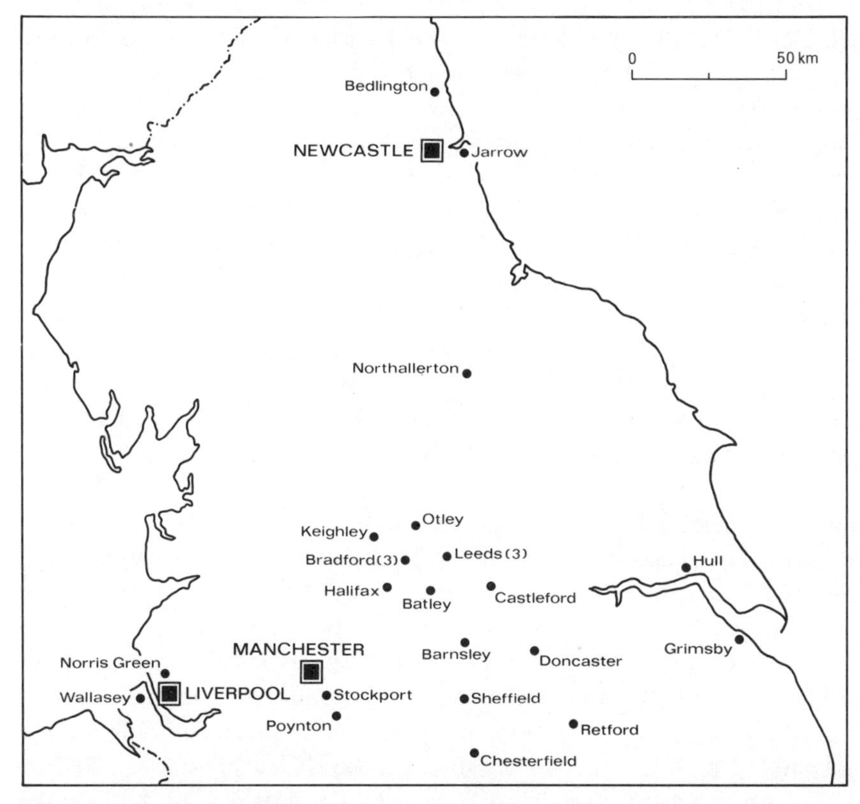

Figure 3.4 Netto's UK store network, February 1992. (*Source:* company press releases.)

reassured when, immediately following its entry into the market, Aldi exploited the rather exposed relations between the food retailers and the regulatory authorities by lodging a formal complaint with the OFT (*The Grocer,* 2 June 1990: 4). Aldi alleged that its right, under EC competition law, of free entry into the British market, was being frustrated by the established UK grocery retailers putting undue pressure on suppliers to cease supplying Aldi. After investigating the allegations, the OFT reported (*The Grocer,* 2 March 1991: 4) that it could find "no firm evidence of concerted or collusive action by the supermarkets or that individual (grocery retailers) had used their buying power in such a way as to constitute an anti-competitive practice", but stated that it would "continue to examine allegations that some manufacturers [had] refused to supply Aldi because of its low pricing policy". However, by that time Aldi had been able to exploit the OFT complaint amongst its customers, and more widely in the press, to reinforce its aggressively low price image. Moreover, it had lodged a second complaint with the

OFT, claiming that suppliers were voluntarily refusing to supply Aldi because they did not like the chain's low-price policy (*The Grocer*, 13 July 1991).

By 1992, two years after Aldi's entry into the UK market, some of the more exaggerated initial fears of its competitive effect had receded. The pace of Aldi's expansion had not been as rapid as first feared. This was partly as a result of certain difficulties over finding sites for expansion. The incumbent retailers undoubtedly engaged in a certain amount of obstruction, pre-emptively buying sites and exerting pressure not to sell, lease or rent to Aldi (Gascoigne 1992, 1993). Labour problems, resulting from the extremely stringent working practices imported from Germany, also contributed to the difficulties. These included manual work for store managers, very few full-time employees, zero-hour contracts in which part-time workers were sent home if trade was slow, and price/product code memorization by checkout staff (Kleinwort Benson 1990, Gascoigne 1992, 1993). In addition, and despite OFT monitoring, Aldi continued to experience difficulties in arranging supplies of leading UK-produced branded products. As a result, Aldi was obliged to stock a higher proportion of less familiar secondary or tertiary UK brands, and unfamiliar Euro-brands or "exclusive labels" imported from its continental suppliers. This allowed some of the incumbent retailers such as Argyll and Tesco (Gascoigne 1992, 1993) to introduce special "price fighter" brands to compete with Aldi, when threatened in specific local markets, without adversely affecting their normal product range and margins elsewhere. Also, it allowed the major UK discounter, Kwik Save, the opportunity to compete when necessary in local markets against Aldi by offering more well known national brands at similar, or slightly higher, prices against Aldi's less familiar brands.

Furthermore, by 1992, it was clear that the wider price-competition spillover effects feared at the time of Aldi's entry into the UK market had been overstated. Aldi had entered a subsector of the market which, to a large extent, was totally separated from that occupied by the major food retailers. Whereas a typical superstore of the major retailers stocked 15,000 product lines, an Aldi store of typical size 560 m^2 (6,000 sq ft) stocked a mere 600–650 lines. As a result, the major retailers could, at one and the same time, contain the competitive pressures of the discounters by holding down gross margins on the standard grocery items, while achieving margin growth in the areas such as fresh produce, chilled foods, and bakery items, in which the discounters could not compete. When added to the fact that the customer profiles of the major retailers and discounters were totally different in terms of social class, income group and access to cars, and to the fact that customers of the major retailers had become accustomed to much higher levels of service than those provided by the extremely low-capital-intensity Aldi-type operation, it became clear that direct price-competition effects were unlikely to cross the market-segment boundaries.

Nevertheless, despite the slower than anticipated pace of expansion and less widespread market effects of the European limited-line discounters, their competitive impact on the UK grocery market was by no means insignificant in the initial phase of their expansion. One market development which their entry did, in part, prompt was the development in the UK of the discount food-warehouse format. This innovation which was copied from the USA – where the "warehouse foods" format had been developed by SuperValue Stores Inc., trading as Cub Foods – emerged in the UK in 1991 in the form of the Food Giant and Pioneer chains. The Food Giants were converted from redundant and grossly underperforming Gateway superstores of average size 3,250 m^2 (35,000 sq ft); the Pioneer stores were converted from the Co-op's less successful Leo superstores of average size 1,860 m^2 (20,000 sq ft). These stores had low residual property values and could be converted at limited cost. Hence, they had a genuine low cost base which facilitated low margins and low prices. Conversion to discount warehouses, with prices matching the Aldi levels, thus allowed a competitive response to the entry of the European discounters by two of the incumbent retailers most directly affected by that entry. At the same time, it facilitated a vital leveraging of low cost assets (County NatWest WoodMac 1991b: 15).

Initial results from the discount food warehouses showed considerable success, and Asda (the other highly pressurized, debt-incumbered, second-tier retailer) began to experiment with a similar format via its Dales discount chain, launched in mid-1992 (*The Grocer*, 20 June 1992, 18 July 1992). A key feature of these developing discount food-warehouse chains, when compared with the Aldi and Netto operations, is their much greater similarity to the superstores of the "big three" in terms of features such as product range. Together with the potentially much larger total turnover of the discount warehouse chains, considered as a group, it follows, therefore, that the wider competitive effects of Aldi and Netto's entry into the UK market might be seen in the form of a significant secondary echo. That is to say, it is possible that the very limited-range, extremely low-margin, Aldi-type discounter might, by the later 1990s, be seen to have triggered the emergence in he UK of discount food-warehousing on a significant scale. Based on the discount warehouse format's success in the USA, and the experience of the impact of such warehouses on the food retailing economies of certain US cities, it is possible that this secondary impact of the entry of the European limited-line discounters might be considerable. Indeed, in terms of its potential effect on selected UK regional or metropolitan grocery markets, the secondary echo might prove to be a formidable amplification of the primary impact.

Conclusions

It is impossible to analyze British retailing – that dynamic and central part of the UK service economy – without being acutely aware that a third phase of the spatial switching of retail capital has been entered. As the international scale of switching unfolds, it has profound consequences for the competitive structure of the industry. In this chapter, two facets of that internationalization have been considered.

Outward movement of UK retail capital has been viewed as a response by the largest grocery retailers to growing doubts about the long-term return on capital from seemingly ever-accelerating UK superstore expansion programmes. In particular, it has been shown that capital-investment opportunities in growth areas are required for an era, in the later 1990s, when expansion opportunities in UK food retailing can be expected to decline, capital-expenditure requirements will fall, but huge cash-flows from the multi-billion turnovers of the top corporations will continue unabated. Sainsbury's expansion into the USA provides, as yet, the best illustration of these themes. Significantly, however, Sainsbury's internationalization has more recently been mirrored by Tesco who, in December 1992, purchased the Catteau chain of 90 supermarkets in northern France for £176 million. Although a relatively small purchase in terms of Tesco's total annual capital budget, the move into mainland Europe has been regarded as immensely significant. It is viewed as indicating the beginning of a major long-term continental European diversification strategy by Tesco, and as signalling slowing growth prospects in the upper end of the UK grocery market. Moreover, after many years in which the corporate trajectories of the two largest UK food retailers have been very similar, the geography of diversification – that is to say the chosen route of internationalization – has suddenly become a critical dimension along which future corporate performance will be assessed.

Inward movement of retail capital into UK food retailing has been seen to be concentrated into the lower-end, discount sector of the market and to come principally from mainland Europe. In contrast to the raising of entry barriers by hugely increased investment and differential access to capital, which characterized the corporate strategies adopted at the upper-end of the market, low capital intensity but high ROCE became the key to market success in the discount sector of the market. This in turn implied low barriers, of both entry and exit, for new competition, and that competition was attracted by envy of high and rising net-profit margins being achieved in UK food retailing and by a sense of the emergence of opportunities for aggressive low-margin operations. It has been seen that some of the initial fears of price-competition ripple effects and destabilization of profit margins which would result from the entry of the European limited-line discounters were rather overstated. Nevertheless, although during the initial phase of European discounters' expansion, direct price-competition effects of the extremely low-margin Aldi-type operations do not appear to have crossed market-segment boundaries, they have prompted a potentially highly significant secondary effect. By triggering the emergence in the UK of discount food-warehousing, and by drawing into that sector the highly pressurized, second-tier retailers, the conditions have been set for much wider price-competitive ripple effects, particularly in certain "value sensitive" urban/industrial regions. In turn, this throws into sharper focus the long-term returns on capital which might be achieved from the major corporations' multi-billion superstore-expansion programmes in those areas, is compounded by the consequences of the overvaluation of their retail property portfolios (Wrigley 1992b), and becomes a significant factor prompting the search for diversification via internationalization. In this way, internationalization at the upper and lower ends of UK food retailing is intrinsically interrelated, and must be viewed as a single process.

Given the centrality of international diversification to the future trajectory of corporate growth in British food retailing, what is surprising, however, is how very little is yet known about the key features of that internationalization. The available literature in geography and planning, indeed the available literature from any source, is remarkably sparse. This chapter has summarized what is currently available and has attempted to provide a framework for future analysis. Two types of research are now required. On the one hand, detailed case studies of market entry and competitive response will provide insight into "the spatial expression and manifestation of retail restructuring" and the manner in which "retail capital penetrates specific spaces" (Ducatel & Blomley 1990: 225). On the other hand, attempts must be made to position the study of the internationalization of British food retailing within broader theoretical debates on the interaction of corporate restructuring and market structures (Clark 1993), and to understand the internationalization of retail capital within a framework which places it as a subform of commercial capital within the larger circuit of total capital (Ducatel & Blomley 1990).

PART TWO

TRANSFORMATION AND THE URBAN REGION

CHAPTER FOUR
The proliferation of the planned shopping centre

Jonathan Reynolds

The transformation of the system of urban shopping centres associated with urban decentralization was first seen in the USA. Planned shopping centres at a variety of scales were commonly developed in accessible locations on the intra-urban highway network. Lord (1988) records that the number of such centres increased between 1950 and 1980 from 100 to 22,000 in a process that has become known as the "malling of America". With steadily increasing levels of car ownership in developed countries it is not surprising that in the 1960s and 1970s in Europe development pressures emerged for similar retail phenomena. The strongest early response was in France, where the 1972–4 boom in planned shopping centre development has been termed a "golden age" by Metton (1983) among others. During this period, nearly half of the present network of peripheral shopping centres in the Île de France region was built. By the mid-1980s, drawing from North American, Australian and French perspectives, Dawson & Lord (1985: 1) felt able to remark that "the shopping centre, over the past thirty years, has become an established feature of urban structure in countries with widely divergent urban policies".

The development of planned, peripheral centres in the UK was, however, slow, because the country had one of the most rigorous and restrictive land-use planning systems. This state of affairs persisted until the mid-1980s: as late as 1985 it was true to say that "the most significant characteristic of British shopping centre development is the fact that nearly all large centre development has taken place in existing town centres" (Schiller 1985: 41). This changed dramatically after 1985. More than 6.6 million m^2 (71 million sq ft) of floorspace was added to UK stock during 1986–92, only 44 per cent of which was in town-centre locations. The rest consisted of out-of-town centres or retail warehouse park developments (Hillier Parker 1991). This rapid expansion was notable in terms of the scale and composition of developments, as well as their location. The first truly regional-scale shopping centres appeared within UK peripheral locations.

The relationship between contemporary UK experience and that of other European countries is not, however, entirely clear. In the light of the increasing internationalization of retailing activity, it is instructive once again to determine whether all European countries are likely to develop similar retail phenomena, despite varying economic, social and institutional contexts. As yet, relatively little comparative work has been undertaken at the pan-European scale to describe and explain the geography of the shopping centre phenomenon (Merenne-Schoumaker 1983, 1991). This issue forms the focus of this chapter. The principal objective is to determine the degree of convergence in the characteristics of certain kinds of shopping centres, which may be the result of a set of common processes in action. The extent to which it might be possible to develop a generic typology of centre types which would transcend national boundaries is also explored.

Definition of the planned shopping centre

Arriving at a common understanding of what is meant by the planned shopping centre across Europe is problematic. Even within the UK, such problems are commonplace: "anyone who has discussed the 'size' of a scheme with a developer knows how hard it is to obtain figures on any common basis" (Thorpe 1992: 3). There is a degree of agreement on the broad nature of a planned shopping centre, although the amount of detail considered relevant by individual sources and the size thresholds adopted vary. Three examples illustrate the point:

- France: "A group of retail outlets usually built to a coherent plan and possessing common components and services (such as parking) . . . The Panorama census is limited to commercial centres larger than 5,000 m² Gross Leasable Area (53,800 sq ft) and/or 10 units." (Panorama 1991: 20).
- UK: "a centre should have at least 4,650 m² GLA (50,000 sq ft), be built and let as an entity and comprise three or more retail units. It should also include some planned pedestrian area outside the component shops or joint car parking facilities." (Hillier Parker 1987: 43).

- Germany: "Shopping centres are centrally planned, built and maintained developments which satisfy the short, middle and long-term requirements (of consumers). They are characterized by:
  - a spatial concentration of specialist non-food, food or service outlets of various sizes;
  - a number of smaller specialist outlets in combination, as a general rule, with one or more dominant operator (such as a Warenhaus, Kaufhaus or SB-Warenhaus);
  - a large shared parking area;
  - a central management;
  - a set of common functions (such as marketing and publicity for example); and – covering a selling area in excess of 10,000 m^2 (107,600 sq ft) or GLA of 15,000 m^2 (161,500 sq ft)" (DHI 1990: 13).

Generally, the simpler the definition, the more kinds of centre may be fitted into the classification. The German definition above is particularly restrictive: a centre without a dominant anchor would not, "as a general rule", be classed as a shopping centre. Most national classifications accommodate three types of planned centre: those that serve "regional" markets (however defined), those that serve a number of community or district needs, and centres designed to cater for local neighbourhood needs. In addition, newer forms of development, such as *parques commerciales* or factory outlet centres, are recognized where they occur in isolation and are "bolted on" to existing typologies. However, the quantitative boundaries for all these forms of development are widely divergent. For example, the threshold in terms of gross leasable area for so-called "regional-scale" centres is 15,000 m^2 (161,500 sq ft) in Germany and Denmark, 20,000 m^2 (215,300 sq ft) in Belgium, 30,000 m^2 (322,900 sq ft) in France and either 38,300 m^2 or 47,800 m^2 (412,300 sq ft or 514,500 sq ft) in the UK, depending upon the authority. There are no guidelines in Spain, Portugal or Italy.

Northern European countries, in which the planned shopping centre is most prominent, tend to maintain the best statistical bases. Government tends to play a relatively small rôle in data collection even here. Consequently there often exist competing definitions and coverage of centre provision in individual countries, generally conducted either by trade associations or market research agencies, or by the local council for shopping centres. By far the most consistent and comprehensive data sets overall exist in France and Germany. (Panorama 1991, DHI 1991). Some of the smaller marketplaces (such as Eire, the Netherlands and Finland) are also well covered (UCD 1990, Ministry of the Environment 1991). Southern European markets are less well served. While it is understood, for example, that there are in the order of 150 shopping centres in Spain, it has only been since the early months of 1992, with the publication of the Corporate Intelligence Group's European Directory of Shopping Centres that there has been some degree of consensus on their size or composition. Even here substantial revisions of data are proving necessary (Corporate Intelligence Group 1992b).

Should figures collected by non-governmental agencies be treated at face value? In the case of floorspace, for example, while the notion of GLA as a unit of measurement is becoming more widely adopted, there is still much reporting of net sales area, building area or site area that makes inter-country comparison difficult. Often, formal definitions are not given. Time-series analysis is also made difficult because of incorrectly reported opening dates and lack of data on incremental additions to floorspace after initial opening. For example, Germany's Ruhrpark, which opened in 1964, had a GLA of some 97,000 m^2 (1.04 million sq ft) in 1992. However, it opened with only 24,000 m^2 (258,300 sq ft), with subsequent additions in 1974 and 1989. A formal time-series analysis should of course allocate this space to the appropriate year.

The bulk of the research reported in this chapter focuses on planned centres larger than 15,000 m^2 GLA (161,500 sq ft). The 15,000 m^2 threshold is a practical one. Accurate reporting of planned centres smaller than this is rare. A 30,000 m^2 (322,900 sq ft) threshold is also used in examining the very largest centres (of a regional scale) across Europe.

Penetration of the planned shopping centre in Europe

Centre distribution

There are significant geographical variations in the penetration of integrated shopping centres in Europe, and in particular of regional shopping centres. Figure 4.1 shows the pan-European distribution of centres of more then 15,000 m^2 GLA (161,500 sq. ft) in 1990 (for countries where meaningful figures are available). The dominance of France, the UK and, to a lesser extent, West Germany are clear. Considering such activity in terms of size of market shows somewhat different results. Figure 4.2 displays shopping-centre development in relation to urban population. A southern European block clearly shows low densities, offering potential for development. Some of these opportunities are

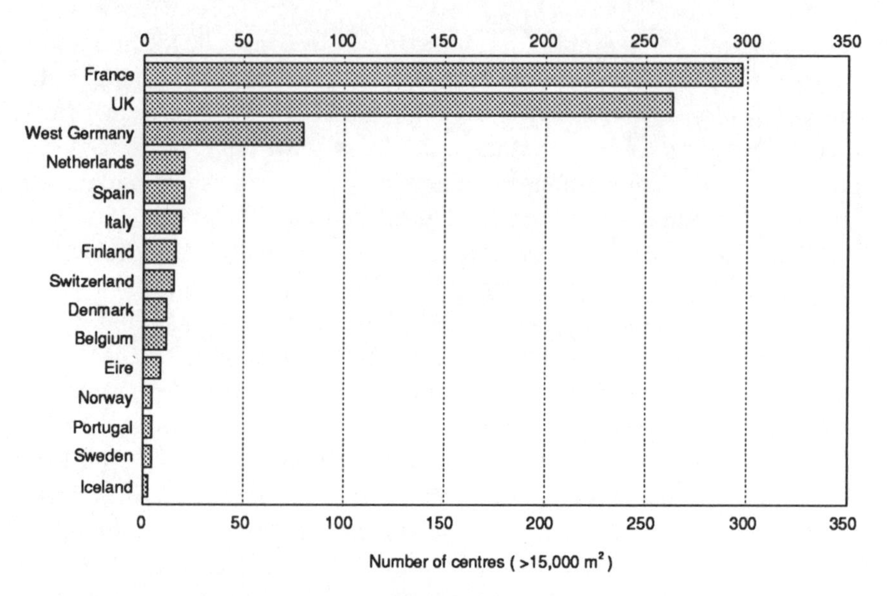

Figure 4.1 European planned shopping centres, 1990.

already being taken up: the Corporate Intelligence Group reports that in Spain alone some 11 schemes of 15,000 m^2 (161,500 sq. ft) or more, totalling 337,000 m^2 (3.63 million sq. ft) were built between 1990 and 1992 (Corporate Intelligence Group 1992b).

Figure 4.2 also shows a contiguous central European block, comprising Denmark, the Netherlands, Belgium, West Germany and Switzerland. These are countries characterized by restrictive planning laws, in general offering a high degree of control of new retail development by existing independent operators and vested interests, as well as by government. In the Netherlands for example, "it is expected that in the next ten years, 3.72 million m^2 (40.04 million sq ft) of retail will shift from weak to strong urban areas" and that "despite the slowdown in growth, developers still have a vast market for the expansion and renovation of numerous (existing) projects" (van Steek, in Roberts & Frampton 1991: 3). France has the highest level of penetration of floorspace per capita within Europe: yet it still manages to add some 20 schemes of 5,000 m^2 GLA (53,800 sq ft) or larger annually to stock, largely provincial or infill Parisian developments, in addition to considerable efforts in upgrading and refurbishing existing centres. The northern European picture is less uniform. There are strong contrasts between the UK experience and that of Sweden and Norway. Finland stands out with a relatively high penetration of large centres, because of its small

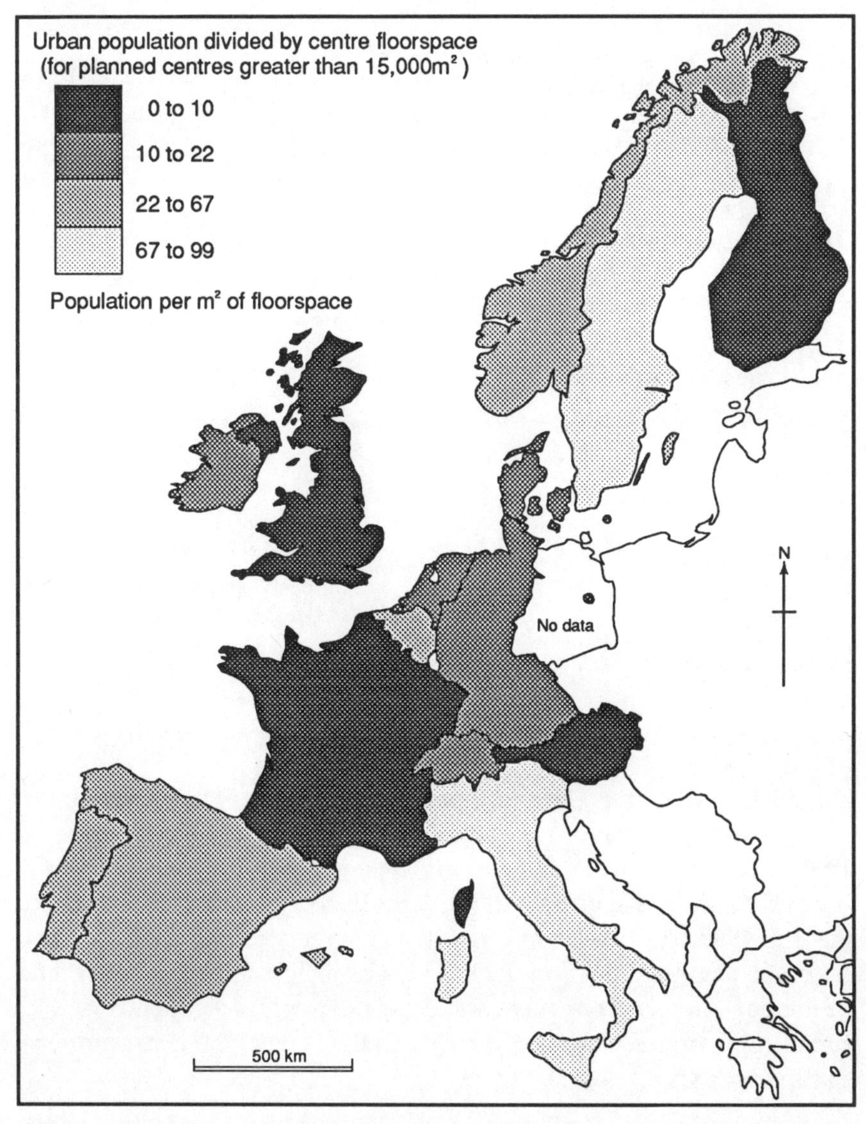

Figure 4.2 European planned shopping centres in relation to urban population, 1990.

population.

By examining the growth of the planned shopping centre across Europe, we can gain further insights into proliferation. Figures 4.3 and 4.4 show data for centres larger than 15,000 m^2 GLA (161,500 sq ft) in a number of countries, for the period 1964–92. It is possible to distinguish three broad sets of patterns, largely determined by market size and other external constraints, many of which relate to the availability of investment. A cyclical model characterizes the development process in

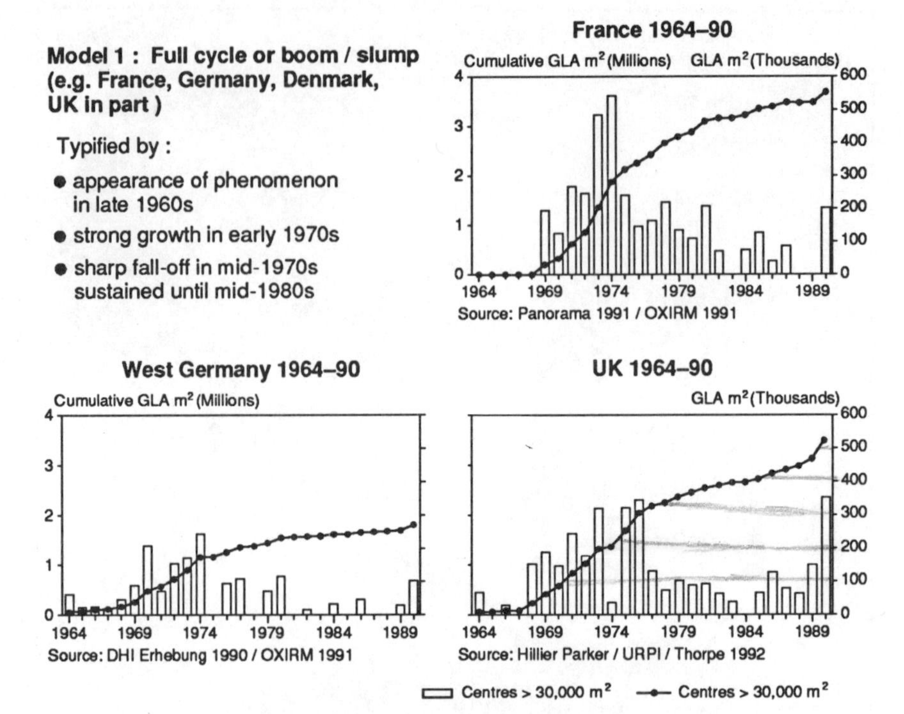

Figure 4.3 Sequence of shopping-centre development in Europe: Model 1.

a number of countries (Fig. 4.3), notably Germany, France and the UK. This boom/slump pattern is typical of the property market (Reynolds 1990). In this case, it heralds an early appearance of the phenomenon in the mid-1960s, sharp growth in the early 1970s and a decline through to the mid-1980s. The UK's more recent activity is a consequence of a less rigorous planning environment and the opportunities retailers and developers expect from increased levels of consumer spending. Not surprisingly, this period has been followed by one of the worst property slumps for many years.

Centre location

There is little reliable published information on the locations of planned centres in relation to established shopping streets across Europe. Nor are there definitions of "in-town", "edge-of-town" or "out-of-town". Some discussion has taken place in individual countries of the impact centres may have on traditional pattern of retailing (DOE 1992a, CNRS 1991). Indeed, such discussion has prompted much of the restrictive

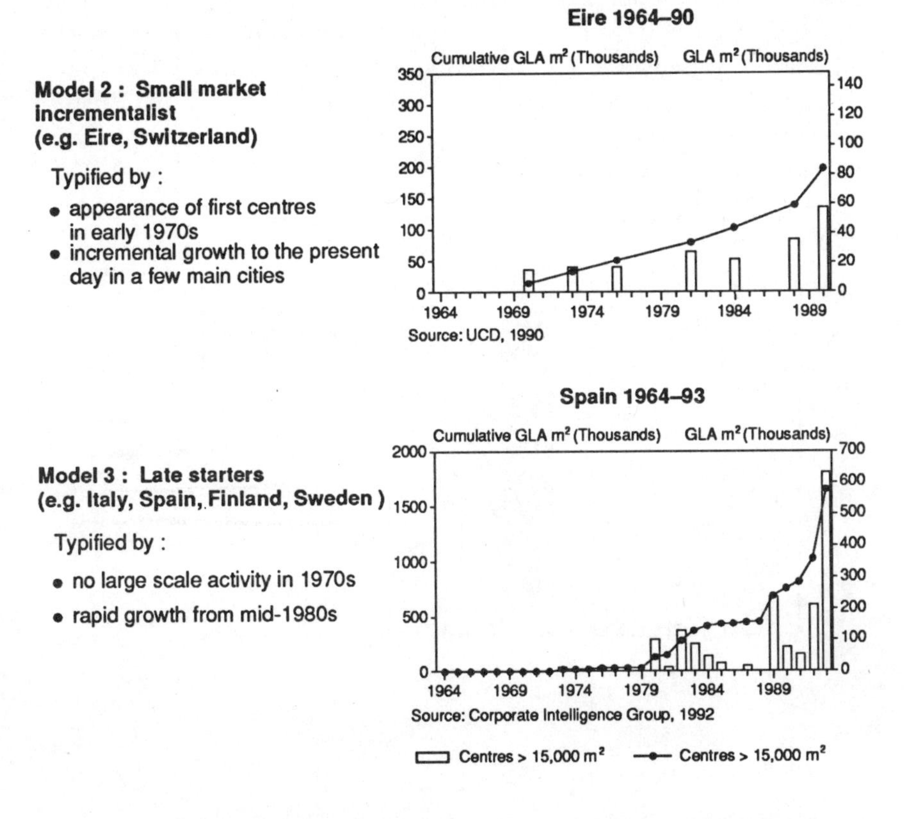

Figure 4.4 Sequence of shopping-centre development in Europe: Models 2 and 3.

planning legislation in many European countries. Thorpe (1992) has discussed definitions of UK shopping centres in terms of town centre, district centre and retail park; German and Finnish analysts have undertaken similar work.

Figure 4.5 summarizes these data for the UK and Germany. Despite recent concern, the bulk of the largest UK centre development has occurred in town-centre locations. The position in Germany is a consequence of four periods of development. Early large-scale peripheral developments were succeeded by intervention that ensured that later, smaller developments took place in or adjacent to central areas. Larger more design-conscious developments during the 1980s have been succeeded by a fourth phase of renewal and revitalization of centres built during the first two phases, and conversion of existing large city-centre stores into speciality centres. In France, Belgium, Spain and

Switzerland, on the other hand, the larger centres tend to be used as growth poles for other urban development. Urban planners have sought to weave in a wide range of social and cultural functions alongside commercial activities (Merenne-Schoumaker 1991). Evry 2 was one of the first French centres to conform to this pattern, while the Spanish Zocos (urban development zones) around Madrid, for example, have new regional centres at their core.

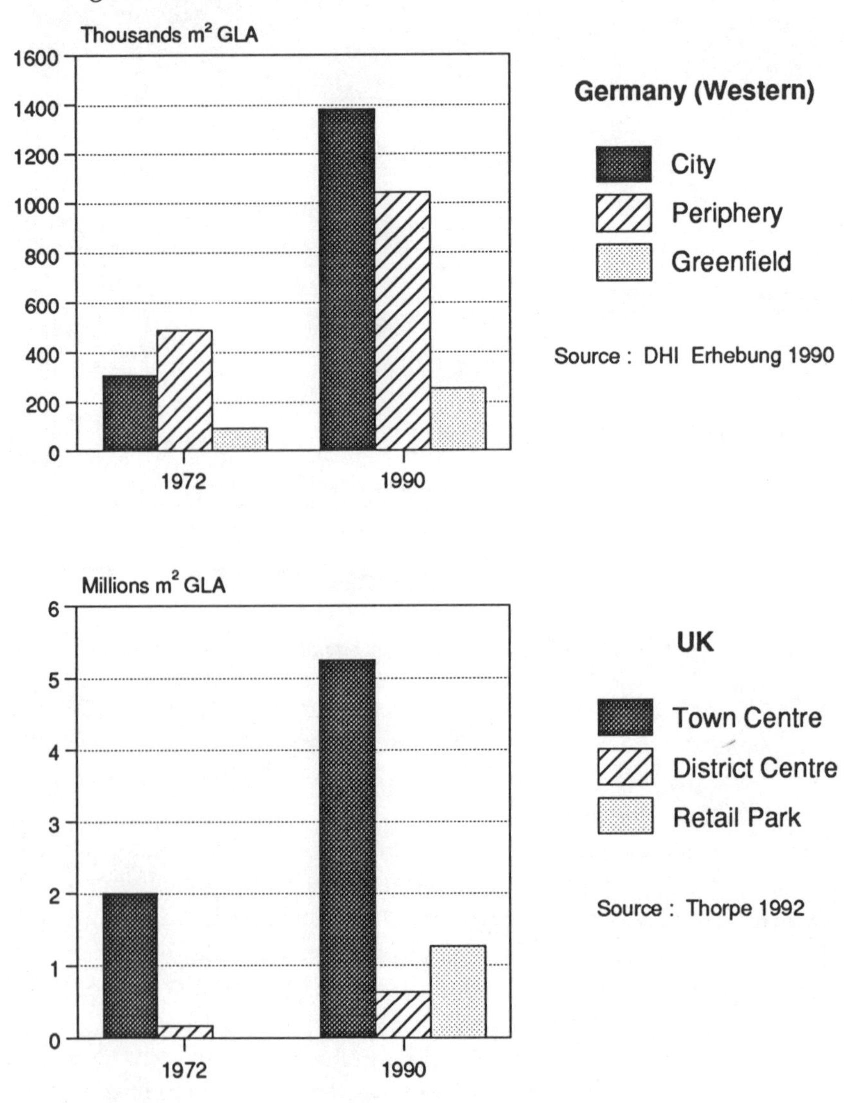

Figure 4.5 Location types of planned shopping centres in West Germany and the UK.

Centre composition

Examination of planned centre composition across Europe provides some of the most significant pointers to the phenomenon's differential penetration within national markets. The US is often cited as the origin of many of the notions that have subsequently manifested themselves as planned centres since the 1960s. Visits paid by French developers to the US in the mid-1960s and of UK entrepreneurs to West Edmonton Mall and other large-scale retail/leisure schemes in the US and Canada during the early 1980s have had direct consequences in Paris, Newcastle upon Tyne and Sheffield.

Yet exclusively European developments can be seen on the ground as well as US development types that have not functioned well in a European setting. The French *centres intercommunaux*, of which there are more than 450 of 5,000 m^2 GLA (53,800 sq ft) or larger, are one example. Anchored by a hypermarket alongside a range of small specialist units, the centres were originally a way for French hypermarket retailers to allow for the continued presence of independent outlets (Delobez 1985). Other anchors have included DIY or furniture and household goods stores. *Centres intercommunaux* have also proved popular in Spain, Portugal and eventually Italy. French hypermarket operators have either directly extended their operations, as into Spain, or as in Italy have undertaken joint ventures with northern Italian retail companies, using, for example, La Rinascente's Città Mercato hypermarkets or Bricocenter DIY stores as anchors.

The other main "home-grown" European development has been the retail warehouse park, which accounted for a great deal of the out-of-town growth in the UK during the late 1980s. In 1992 it was estimated that there were nearly 250 parks trading in the UK (Staniland Hall 1992). *Fachmarkte* in small clusters are relatively common in Germany, but parks are relatively uncommon in France. A small number of *centres de magasins d'usine* are related phenomena, but have not been particularly successful (Knee 1988).

Even the conventional US regional centre model has not always translated well into the European context. In France, regional shopping centres in the Île de France one by one replaced the department store anchors suggested by conventional wisdom with hypermarkets or speciality stores such as Grand'Habitat (Knee 1988). Retailers argued that consumers still preferred to use traditional Parisian shopping streets for fashion clothing purchases. A direct consequence of this was the forced closure of the smaller supermarkets already represented in the regional centres. Planning the leisure component within the UK regional centres of the late 1980s in line with US principles was also fraught with difficulty. The extent to which leisure would work in tandem with such developments was doubtful; in the end, substantial scaling down of leisure in Gateshead's MetroCentre took place - and leisure as such failed to appear at all in Sheffield's Meadowhall Centre or Thurrock's Lakeside.

A typology of planned shopping centre development

An attempt can be made to reconcile the confusing and somewhat conflicting classifications of shopping-centre development in individual countries into a more generic form. Previous experiments at the national level have already been made by academics and practitioners, but these have included unplanned as well as planned forms of development (for example, Heineberg & Mayr 1988, Reynolds & Schiller 1992). There has also been discussion of the extent to which present-day patterns of retail activity still contain a hierarchical component (Dawson & Sparks 1987). In 1990, Brown suggested that three basic variants of physical form and function can be combined to provide "a non-hierarchical matrix of locational archetypes" (Brown 1990). This chapter is concerned with one subset of retail locational types, those that Brown calls "planned clusters".

Table 4.1 provides a tentative framework, derived from observed characteristics of European planned shopping centres. Variants are distinguished that are either associated with a centre's location (its degree of centrality and relation to other activities) or with its composition (the nature of the tenant mix and kind of anchor store). This typology permits sufficient variants for 19 differentiated types of development, or niches. Aggregating the 19 detailed types allows any centre to be further identified with one of four broad types of centre: regional, intermediate, park, and speciality. These are ultimately differentiated in terms of three factors: their GLA, the extent of scheme integration, and the numbers of anchors.

Of course, not all the detailed types may be present within every country in the analysis. Intermediate centres are most widely known through France's *centres intercommunaux* and are so far relatively new features in the northern Italian market and in the joint Marks & Spencer/Tesco or Sainsbury "sub-regional" developments in the UK. There are therefore a number of vacant niches in many countries that

Table 4.1 Typology of planned shopping centres in Europe.

*I: Regional shopping centre (30,000 m^2+; 322,900 sq ft)**		
(Centres commerciaux régionaux, grandes centros periféricos, Regionalen Shopping-Center; two or more anchors)		
Locational variants	– central area in traditional core	Eldon Square, Newcastle, UK
	– central area adjacent traditional core	La Part-Dieu, Lyon, France
	– non-central suburban growth pole	Vélizy 2, Versailles, France
	– green field site/transport node	Curno, Bergame, Italy
Compositional variants	– hypermarket-dominated	A6, Jönköping, Sweden
	– department & variety-store dominated	Lakeside, Thurrock, UK
	– food, non-food and leisure anchors	Parquesur, Madrid, Spain
II: Intermediate centres (10,000–30,000 m^2; 107,600–322,900 sq ft)		
(Centres intercommunaux, centros intermedios; at least one anchor, integrated)		
Locational variants	– on-central suburban community	Auchan, Torino, Italy
	– greenfield site/transport node	Cameron Toll, Edinburgh, UK
Compositional variants	– hypermarket-anchored	Euromarché
	– speciality non-food anchored	BHV, Cergy, France
III: Retail parks (5,000–20,000 m^2; 53,800–215,300 sq ft)		
(Centres de magasins d'usine ou parc des entrepots, parques comerciales, retail warehouse parks; not obviously anchored; not wholly integrated centres)		
Locational variants	– non-central suburban community	various, UK
	– greenfield site/transport node	Lakeside Retail Park, UK
Compositional variants	– retail warehouse tenant mix	Fairacres Retail Park, Abingdon, UK
	– factory outlet tenant mix	Direct Usines, Nancy, France
	– hybrid tenant mix	Fosse Park, Leicester, UK
IV: Speciality centres (1,000 m^2+; 18,800 sq ft)		
(arcades, galeries marchandes, galerías comerciales, Galerien)		
Locational variants	– central area in traditional core	Arcades, Lille, France
Compositional variants	– non-food specialist traders	Powerscourt Town House, Dublin, Eire
	– department store conversion	Centre Point, Braunschweig, Germany

Notes
* Floorspace figures are indicative only.
Centres providing for local or neighbourhood needs are excluded.

are yet to be occupied. There will certainly also exist mixes of locational and compositional attributes that are yet to appear at all within western Europe: the regional scale, integrated factory outlet mall, for example, food service-oriented parks or out-of-town speciality centres (antiques malls, for example, are common in the US).

The operational environment

This chapter can only go a short way in attempting to draw out explanations for differing characteristics of planned shopping-centre development across Europe. Nevertheless it can make some passing observations on some of the most important sets of influences: those that are largely operational in origin (related to retail structure and trading characteristics) and those relating to the planning and institutional environment.

A close link exists between the likely success of planned centres and the degree of concentration in the retail industry. Few independent traders are likely to be able to afford rental levels in planned centres – even if located on cheaper edge-of-town sites. Figure 4.6 shows the penetration of multiple trading across Europe. The poorer penetration of the southern European market by regional-scale planned shopping centres, for example, is almost entirely related to the relatively less concentrated nature of retail distribution in those countries (itself a consequence of a broad range of social and economic characteristics of the population). With a relative lack of strong national or regional retail brands, early proposals for new centres in southern European countries were also likely to founder as a result of the absence of effective anchor operations (Rossi 1990). "International investors like to know who the traders are going to be" (Percival 1991).

This carries with it the implication that the rapid growth of integrated shopping centres is likely to have a consolidating rôle in the evolution of a country's retail structure. New centres are likely to encourage, other things being equal, the growth of standardized retail operations. Martinez (1991) suggests that this is presently occurring in Spain. Metton (1983) comments on the similar rôle that French centres *commerciaux régionaux* played in the Paris region as "tools of commercial concentration".

Developers of two of the major regional out-of-town centres in the UK opened in the late 1980s built into their letting agreements with retailers a clause requiring them to open on Sundays - in anticipation of changes in restrictive legislation. It is certainly true that the emerging leisure shopping rôle of the larger of the new centres is ideally suited to seven-day and late-evening trading. But this is one area where much of Europe lags behind customer demand and where legislation dates back in some cases more than 30 years.

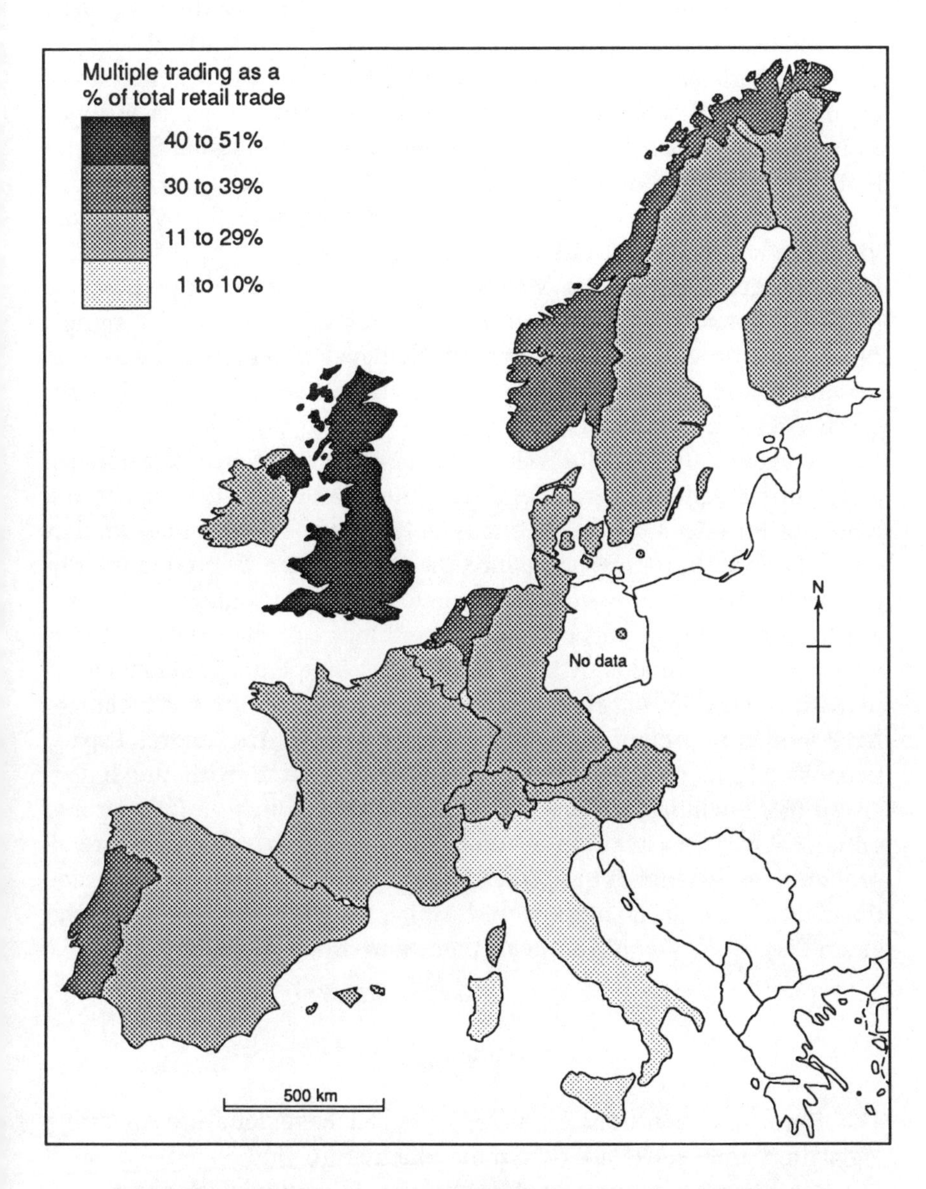

Figure 4.6 Multiple trading as a percentage of total retail trade in Europe.

The planning and institutional environment

The developer in Europe

While the very largest centres have been essentially developer-led (in some cases offering an integrated package from funding through construction to management, as in the case of French developer Ségécé), charismatic individuals have played as large a part in determining the scale and nature of shopping-centre development across Europe, as have anonymous funding institutions. It was Jean-Louis Solal who first brought much of the US experience in shopping-centre development to Europe in the mid-1960s, persuading suspicious French retailers to take space in Parly 2 in Paris. In the mid-1980s, the success of John Hall, the developer of the MetroCentre at Gateshead, in persuading Marks & Spencer to participate in the scheme was of similar significance. This provided the impetus for the development of a major new regional shopping centre, rather than a hypermarket and retail warehouse complex (Schiller 1986).

Convergence in centre development is encouraged by the extensive cross-border activity of retailers, developers and investors across the whole of Europe. Only in Germany is there a relatively small foreign developer/investor presence (since the marketplace is predominantly freehold). There is powerful French involvement in southern Europe – particularly in Spain and Portugal, where many of the major and best opportunities have already been identified. While French retailers have tended to establish subsidiary operations in these countries, developers have tended to prefer joint ventures. For example, Arc Union's Espace Expansion arm established the SdS joint operation with the Italian group Ifile to embark on major growth in northern Italy. Joint ventures with local partners who can obtain prime sites but who have little retail property experience offer considerable potential for both parties (Percival 1991). It is these vehicles that provide the basis for the emergence of common European themes in centre development.

Planning regimes

The European countries discussed here all have land-use planning legislation that seeks to determine the nature and extent of retail investment and the activities of developers. Planning legislation ranges from floorspace- and location-specific measures to zoning and operational controls. Floorspace-specific legislation, such as the Loi Cadenas in Belgium, the Loi Royer in France or the Baunutzungsverordnung in Germany, can in some cases regulate the development of units as small as 750 m^2 (8,000 sq ft). Greek legislation prevents developers building on more than 10 per cent of any site purchase, effectively driving up land prices. Centre growth may also often be stunted through anti-concentration measures. Until 1980 in Denmark legislation controlled the maximum number of branches that could be opened by any one retailer. The Netherlands supports a combined approach in which zoning restrictions and limitations placed upon building height are linked with legislation that tends to support new centres only as part of comprehensive development.

The powerful lobbying position of independent retail operators though local and regional chambers of commerce lies at the heart of much existing legislation. The gradual consolidation of the retail sector is shifting the status quo, however. Changing consumer demand is at work, as is the recognition by local planning authorities that substantial tax revenues can be obtained from large new centres: "it used to take ten years to open a shopping centre in Italy; it only takes five today" (Ferrarini, in Tonion 1991).

Not all restrictions on planned centre development have been wholly successful. Ministerial directives of the early 1980s in Eire sought to avoid urban deterioration through out-of-town development. Nevertheless, between 1981 and 1989 more than 120,000 m^2 GLA of floorspace (1.29 million sq ft) was added in suburban Dublin alone (UCD 1990). In France, although the Loi Royer in principle severely regulates the number of developments in excess of 1,200 m^2 GLA (12,900 sq ft), it has been largely circumvented by loosely affiliated operators such as Intermarché and it is suggested that on appeal the operator has a better than even chance of obtaining planning permission. Indeed, of the 1.9 million m^2 (20.45 million sq ft) of floorspace for which permission was sought in 1987, 1.3 million m^2 – more than two-thirds – was authorized.

The main consequence of planning restraint is the pressure that it exerts upon existing land and developments and upon rental levels, making redevelopment prohibitively expensive in any case. Where this occurs, new centres are the exception rather than the rule. The Ville 2 development, which opened in 1990 at Charleroi in Belgium, attracted attention far out of proportion to its small size, for example. In both Finland and Sweden since the late 1980s, the pressure for development has become so great that planning controls have been relaxed to make suburban and out-of-town locations permissible. The 45,000 m^2 GLA (484,400 sq ft) A6 centre at Jönköping, midway between Malmö and Stockholm, was the first result. Two hypermarket anchors, an Ikea store and 40 specialist shops were constructed in 1990. Three similar centres are planned for Göteburg, Helsingborg and Örebro.

Conclusions

This chapter has sought to analyze the proliferation of the planned shopping centre as it has developed across Europe since the mid-1960s in the context of widely varying operational and institutional environments. Despite the pace and scale of such activity, some of the most basic characteristics of such developments within some countries remain largely undocumented and there is much disagreement about appropriate definitions. The chapter has questioned the extent to which some, even official, sources of information can be given credence.

In the absence of a common statistical base and despite conflicting definitions, it has nevertheless proved possible to make some general observations about the penetration of the planned shopping centre in Europe. The common perception that a southern European block of countries remains relatively untouched by planned centre development, for example, has been challenged and the extent of proposed development in Spain and, to a lesser extent, Italy has been explored. While out-of-centre retailing activity in nearly all European countries is generally higher now than in the past, paths taken have been very different. A helpful way of coming to terms with the differential penetration of centres is to track centre development over time and, as a consequence, three distinct "models" of centre development can be discerned that begin to indicate that the notion of "convergence" in centre development actually conceals considerable complexity. This complexity is heightened when the composition of centres is examined. Despite the common assumption that European experiences in peripheral shopping-centre development are likely to be largely in line with those of the USA, it appears that a number of US development types have failed to work in the European context and that a number of "home-grown" development types have emerged that are not part of the US experience.

In drawing these ideas together, a flexible typology of planned shopping-centre development has been proposed that is derived from the literature's concern with combinations of physical form and function in describing the "planned clustering" of retail activity. The typology permits the successful incorporation of a wide range of development types and can be tailored for individual national markets. Such a typology forms a basis for explanations of the differential growth and character of the planned shopping centre, in terms of both operational and institutional factors. It also reinforces the view that while a broad degree of convergence in retail commercial development activity can be identified, the extent to which retail phenomena are manifested within different European countries reveals a considerably more sophisticated set of processes that both developer and analyst ignore at their peril.

CHAPTER FIVE

Transformation and the city centre

Clifford M. Guy & J. Dennis Lord

This chapter considers the changing rôle of the city centre in the urban economy, concentrating particularly on retail change and its relationships with the local economy and local politics. Use of the term city implies settlements of medium size within the urban hierarchy: of regional significance but below the level of the conurbation. The analysis takes the form of a case study, comparing the centres of two cities of similar size but in different economic and political environments.

In most developed economies, the city centre has been under threat of competition from new retail developments in suburban locations. The main reason is arguably the growth of suburban population, which has generated a demand for retail facilities close to home. The rise in car use for shopping has disadvantaged the central area, where high land values mean that free car parking cannot easily be provided for the customer and traffic congestion can make access difficult. Also, land for retail development is generally much cheaper in the suburbs than in the city centre and sufficient space is available for the very large sites required by developers of new retail space, whether free-standing, single-level stores or enclosed "regional" shopping malls. Retail firms selling goods that shoppers habitually buy in bulk, using their cars, prefer to develop away from city centres. Thus the decentralization of both convenience shopping and "bulky" household-goods shopping is common in North America and many western European countries.

Retail change in city centres has taken two broad paths in response to increasing pressure for suburbanization of retailing. The first has been a steady decline, with major department stores closing down, often having opened new branches in suburban centres. Other shop closures have followed, so that central-area retailing has become reduced to the status of a convenience centre for office workers and visitors to the city centre. This has typified many US cities (Baerwald 1989, Buckwalter 1989, Lord 1988, Morrill 1987, Robertson 1983, 1990) and to some extent Canada (Hallsworth 1988). The second path has been for the city centre to maintain its status as the premier retail area in the city, usually through a combination of new development and environmental improvements, and restriction on major retail development elsewhere. Many western European countries have followed this path (Davies & Champion 1983, Guy 1980, Knee 1988, TEST 1989). In the UK in particular, the fully enclosed shopping mall, developed in North America in suburban locations, has been brought into the city centre to provide high-quality space for comparison-goods retailers.

In both North America and western Europe, city centres have had to adapt their retail characteristics to new markets. Their rôle in serving the day-to-day requirements of shoppers has altered: instead of serving the convenience needs of the locally resident population, they increasingly serve the needs of the city centre's working population. In many city centres, a substantial transient population of tourists and other recreational visitors is also there to be served. This has led to the development of "festival shopping" projects in several cities. In western Europe, festival shopping is relatively unimportant compared with the solid provision of mainstream comparison retailing in city centres. Indeed, festival shopping has often been located in waterfront or other inner-urban locations situated some distance from the city centre. In North America, by contrast, festival shopping has been promoted as the saviour of the city centre, although with variable success (Guskind & Pierce 1988) and often at much cost to the public purse (Sawicki 1989).

This chapter examines these issues through a comparison of recent developments in central area retailing in Cardiff, South Wales (UK) and Charlotte, North Carolina (USA). The reasons for using these cities for comparative purposes have been discussed in previous papers (Lord & Guy 1991, Guy & Lord 1991). Briefly, the two cities have similar population sizes (about 300,000), with comparable retail catchment populations (about 1.25 million), and both are regional administrative and financial centres. There is no pretence that the two cities are entirely "typical" of medium-sized cities in either country. Indeed, no cities can be so. However, they do exemplify the spatial effects of broad trends that have affected retailing in the UK and USA over the past few decades. They are both large enough to possess virtually all of the typical forms of retail development that have occurred in the two countries during this period. From detailed comparisons of the two cities it has also been possible to clarify and extend notions about more general differences between processes of retailing and retail development in the two countries.

The account proceeds in the following way. First, the retail structures of the two city centres as they were in the early 1970s is examined. Attention focuses on similarities in overall size and importance, and in the composition of retailing in the centres. Changes over the past 20 years are then discussed: initially in retail size and composition, and then in relation to locational patterns of retailing. These changes are explained mainly in terms of the series of property (re)development schemes that have taken place. Finally, these changes are related to broader trends in retailing in the UK and USA, and to the general issues discussed above.

The two city centres in the early 1970s

In the early 1970s, both city centres had substantial retail areas: about 241,500 m^2 (2.6 million sq ft) of gross retail floor area in Cardiff, and 148,600 m^2 (1.6 million sq ft) in Charlotte. Charlotte, however, had far fewer retail outlets in its city centre (118) than did Cardiff (463) (Fig. 5.1). It should be noted, though, that differences in definitions of retail categories, and possibly floor area, make detailed comparisons hazardous. A major difference between US and UK definitions of retailing is that the US data normally include automobile sales, auto service stations and restaurants. These categories are omitted from the Charlotte estimates in the comparisons made in this chapter because of the problems involved in including them in the Cardiff data.

Both centres were important mainly for comparison goods, particularly Charlotte. Each centre was anchored by three substantial department stores – Howells, David Morgan, and Evan Roberts in Cardiff, and Belk, Sears and Roebuck, and Ivey's in Charlotte. Charlotte was, however, far more dependent on its three department stores (52 per cent of its retail floor area) than was Cardiff, whose three large department stores probably took up no more than 25 per cent of the city centre's total floor area (exact figures are not available). Cardiff also had another, smaller department store (Seccombe's) and four variety stores owned by national chains (Marks & Spencer, British Home Stores, Littlewoods and Woolworths). These eight department and variety stores used about 46 per cent of Cardiff's net retail floorspace (gross floor-area estimates are not available).

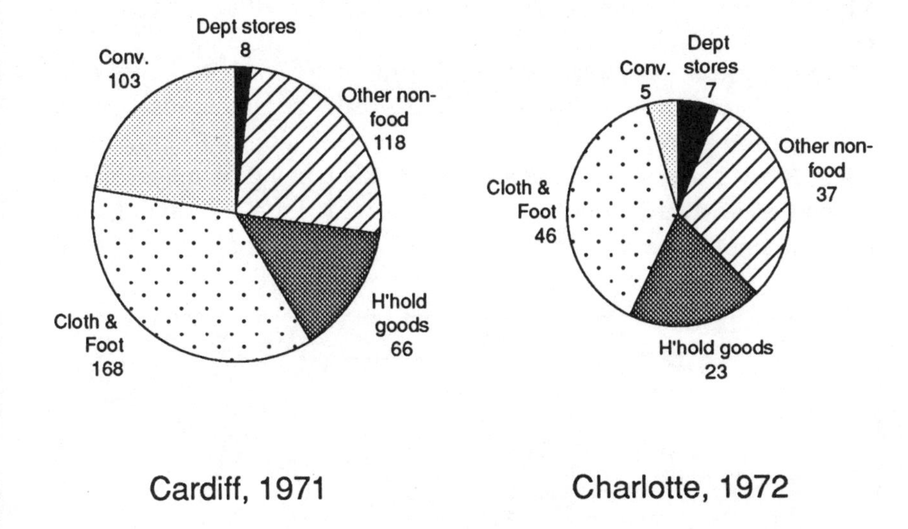

Figure 5.1 Shop numbers in Cardiff and Charlotte city centres in 1971/72.

A comparison of retail turnover data (Table 5.1) suggests that Charlotte city centre's rôle in the city's retail system was much less important than Cardiff's, although it was still the most important shopping centre in the metropolitan area. In fact, Charlotte's central area

Table 5.1 Central areas' shares of retail turnover in Cardiff (1971) and Charlotte (1972) (%).

Type of retailing	Cardiff	Charlotte
"GAF"[a]	70	25
Other	20	11
ALL RETAIL SALES[b]	45	15

Sources: UK Census of Distribution (1971); US Bureau of the Census, Census of Retail Trade (1972).

[a] General merchandise, apparel and furniture. The Cardiff percentages are for the UK Census categories clothing and footwear, household goods and general stores.

[b] Charlotte data include some categories omitted from the Cardiff data, such as sales of cars, petrol, and food and drink in restaurants.

was in a state of decline from the early 1960s, before which it had enjoyed shares of retail trade comparable with Cardiff's city centre (Lord & Guy 1991).

Maps of the two central areas at this time show some contrasts (Figs 5.2, 5.3). Cardiff's shopping area was basically a continuous L-shape with one arm along Queen Street and the other along Working Street and St Mary's Street, with the area inside the L being occupied mainly by wholesale distributors and car parks. Charlotte's shopping area included a notable concentration of retailing around the intersection of

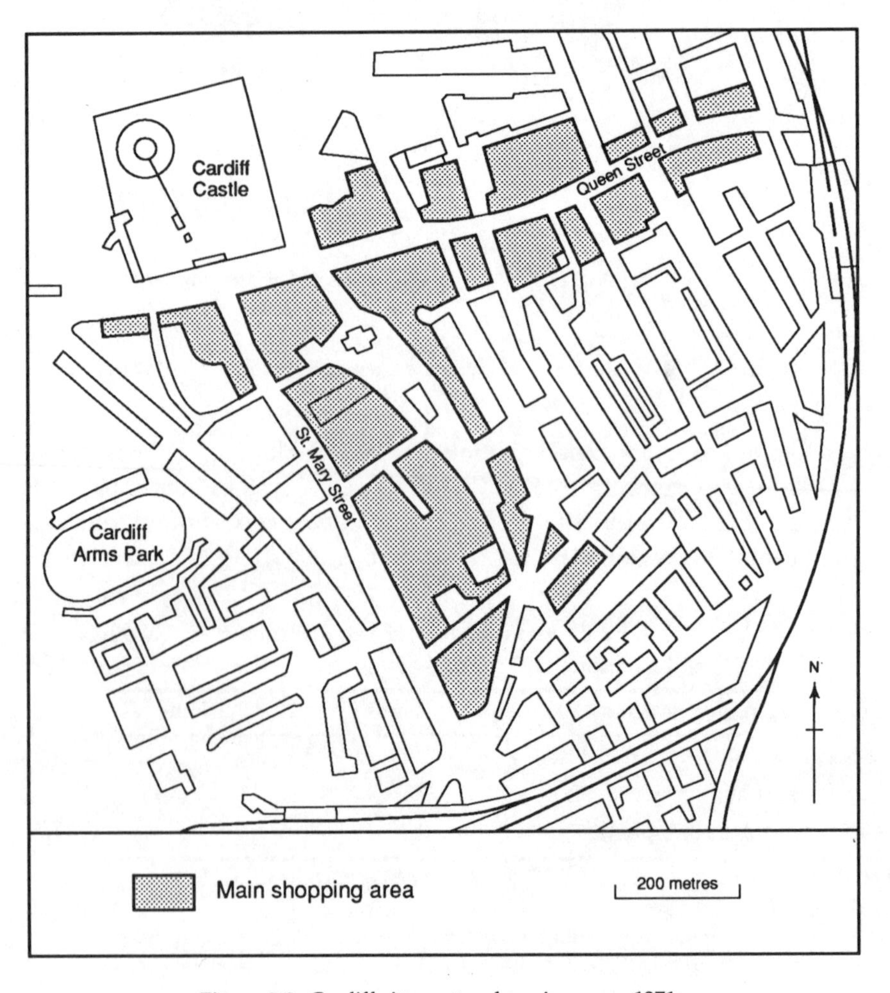

Figure 5.2 Cardiff city-centre shopping area, 1971.

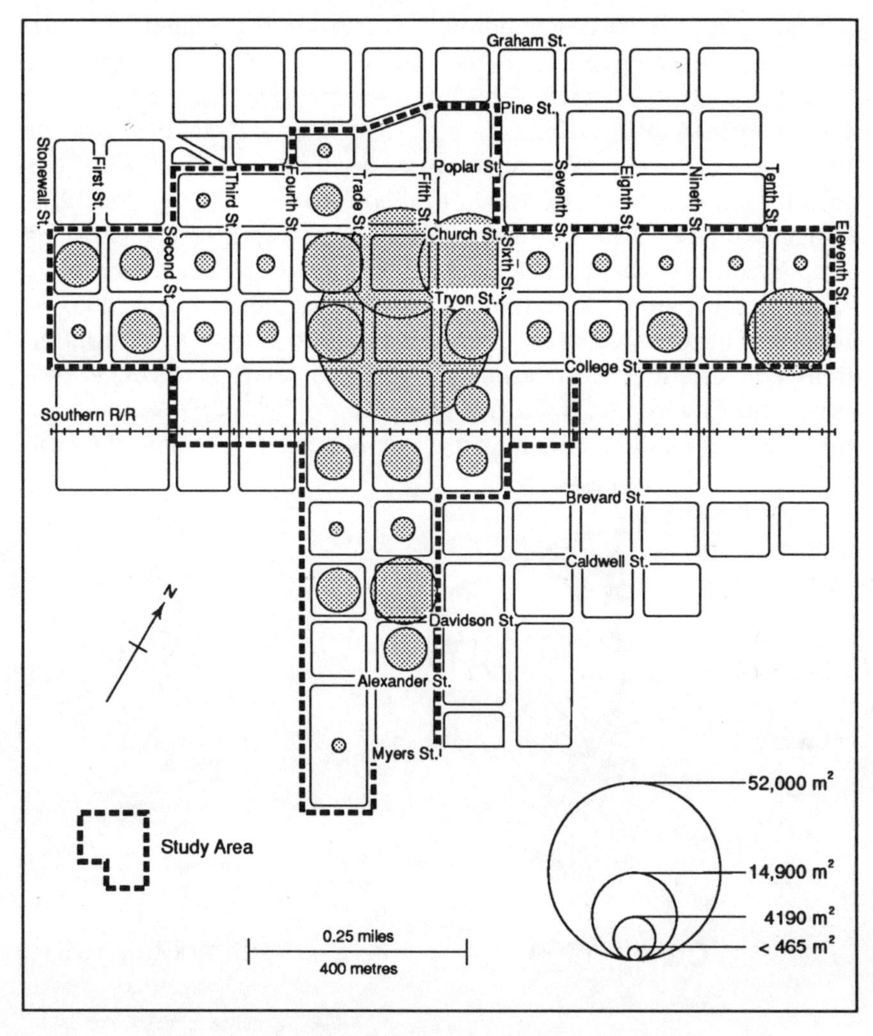

Figure 5.3 Charlotte city-centre retail floor area (m^2), 1972.

Tryon and Fifth Streets. The other main retail site was the Sears store at the northeastern edge of the centre.

Changes in retail size and composition in the 1970s and 1980s

Both centres have been surveyed recently and it is thus possible to compare changes in the volume and type of retailing over approximately the past 20 years. A brief description is also given of the retail decentralization in the same period.

Cardiff centre's gross retail area increased from about 239,600 m^2 (2.58 million sq ft) in 1971 to 253,600 m^2 (2.73 million sq ft) in 1990 (City of Cardiff 1991). Charlotte's, by contrast, declined from 148,600 m^2 (1.6 million sq ft) to only 22,300 m^2 (0.24 million sq ft) over the period 1972–91 (Fig. 5.4). Table 5.2 and Figure 5.5 show changes in retail size and composition over this period. As detailed gross floor area data are not available for Cardiff, the change in floorspace there has had to be measured in terms of sales area. This tends to exaggerate the differences between the two cities so far as gross retail provision is concerned: sales area in Cardiff has risen more quickly than gross area, implying that retail space is now used more efficiently than in the past.

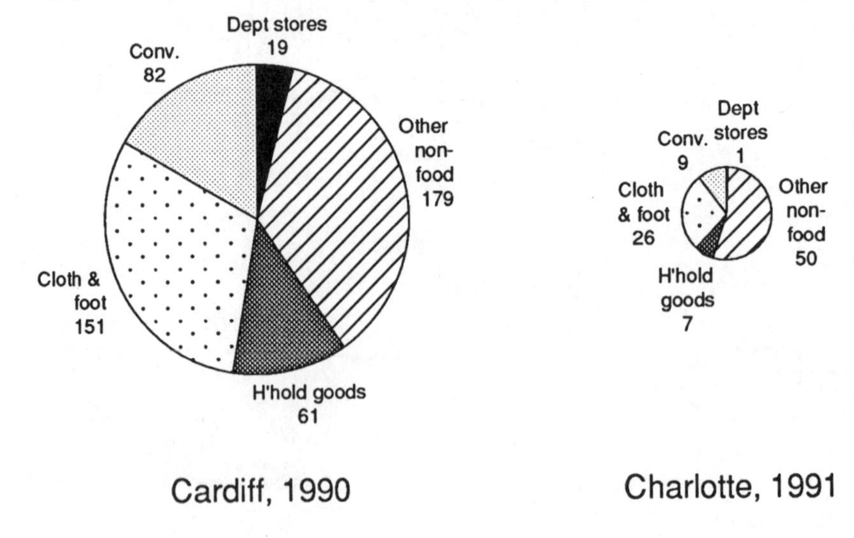

Figure 5.4 Shop numbers in Cardiff and Charlotte city centres in 1990/91.

Nevertheless, a massive contrast exists between the two centres: modest increases in both number of shops and total floorspace in Cardiff, and a massive decline in floorspace in Charlotte, with a smaller decline in the number of shops. Overall, Cardiff has increased its central-area retail floorspace by 5 per cent (gross area) and 24 per cent (sales area), whereas the gross area in Charlotte has decreased by 85 per cent.

Because of the absence of recent turnover estimates for Cardiff and the discontinuation of the Major Retail Center reports by the US Bureau of the Census, data such as that in Table 5.1 cannot be assembled for

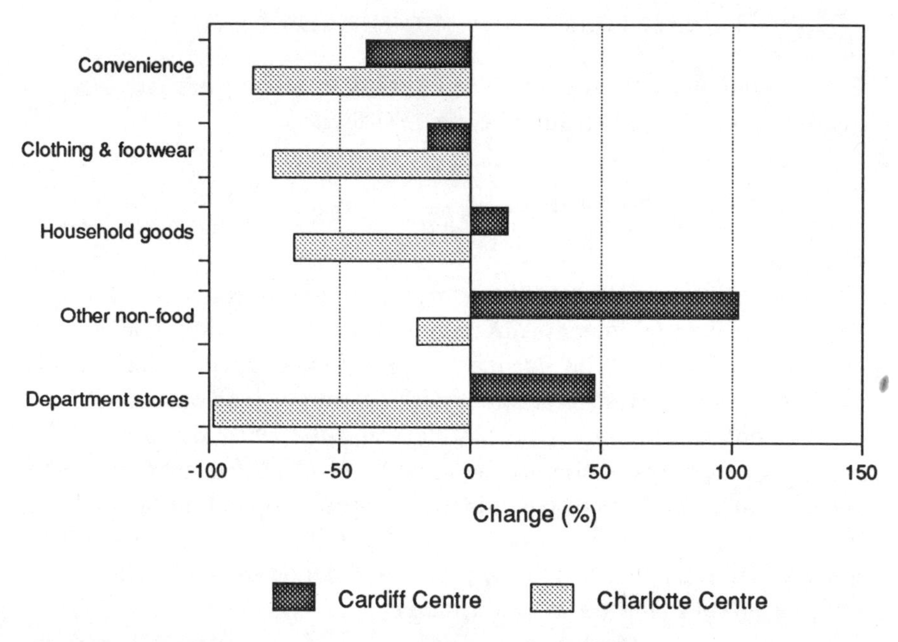

Figure 5.5 Percentage changes in retail floorspace in Cardiff and Charlotte city centres since 1971/72.

Table 5.2 Absolute changes in city centre retail activity in Cardiff (1971–90) and Charlotte (1972–91).

	Cardiff		Charlotte	
Retail category[c]	a	b	a	b
Convenience (54)	−21	−2.0	4	−3.3
Clothing and footwear (56)	−17	−4.6	−20	−17.3
Household goods stores (52, 57)	−5	1.4	−16	−17.2
Other non-food stores (59, 591, 5944, 5992)	61	8.2	13	−2.0
Department, variety stores (53)	11	20.5	−6	−86.2
Total comparison	50	25.4	−29	−123.5
Total retail	29	23.4	−25	−125.8

Sources: UK Census of Distribution (1971); Charlotte Retailing Guide, September 1972; South Glamorgan County Council (1990); Field survey in Charlotte, 1991.
a: Change in number of shops.
b: Change in net retail floor area (Cardiff), and gross area (Charlotte), in thousand m^2.
[c] The categories used are derived from the UK Census of Distribution (1971). They correspond only approximately with the American SIC Codes, which are shown in brackets.

the early 1990s. It is known, however, that Charlotte city centre's percentage of both "GAF" and "other" retail sales in the city as a whole is now well below 5 per cent (Lord & Guy 1991). Cardiff city centre's share is probably only a little below 1971 levels.

Cardiff

The main changes in Cardiff have been a decline in convenience retailing and substantial increases in "other non-food" retailing and department/variety stores. The decline in convenience retailing has mainly taken the form of closure of supermarkets, and of food sections within department and variety stores. This coincided with the opening of seven large grocery stores in the suburbs of Cardiff during the 1980s (Lord & Guy 1991). As discussed below, a Tesco superstore opened and also closed in the city centre during the 1980s. The most significant food outlets in the city centre are now probably the food section in the Marks & Spencer store and the covered market.

The increase in "other non-food" retailing has been affected by growth in specialist and luxury goods stores, particularly in the 1980s. The increase in department/variety store space has several causes, including the opening of the 13,900 m^2 (150,000 sq ft) Debenhams store in 1982, the expansion of the formerly specialist Boots and W. H. Smith companies into mixed-goods retailers in the 1970s, and expansion of two department stores and several variety stores in the 1980s through horizontal extension and/or more efficient usage of upper floors. The figures also conceal the closure of the Evan Roberts and Seccombes department stores during the 1970s, and the Allders store in 1986.

Closures of furniture and carpet stores in the city centre have reflected the growth in off-centre stores of this type. Some 13 stores of more than 1,860 m^2 sales area (20,000 sq ft), selling DIY, furniture or carpets were opened in suburban locations in Cardiff during the 1980s (South Glamorgan County Council 1990, see also Lord & Guy 1991). A recent event of major importance in the Cardiff area has been the opening of a new 8,370 m^2 (gross area) (90,000 sq ft) Marks & Spencer store on the western edge of the city in early 1992. This has not, however, led to the closure of the firm's city-centre store, nor does this appear likely in the foreseeable future. Rather, the wish to develop the new store was based partly upon the city-centre store over-trading and being unable to expand on its existing site.

Charlotte

The main feature of change in Charlotte has been the closure of the three major department stores that did the bulk of retailing in 1972. Losses in floorspace have occurred in all other categories of retailing. The number of shops has decreased much less than has floorspace and numbers have actually increased in some categories. Clearly the new shop units have tended to be much smaller than those they have replaced. This has not been so in Cardiff except in the now very small convenience retail sector.

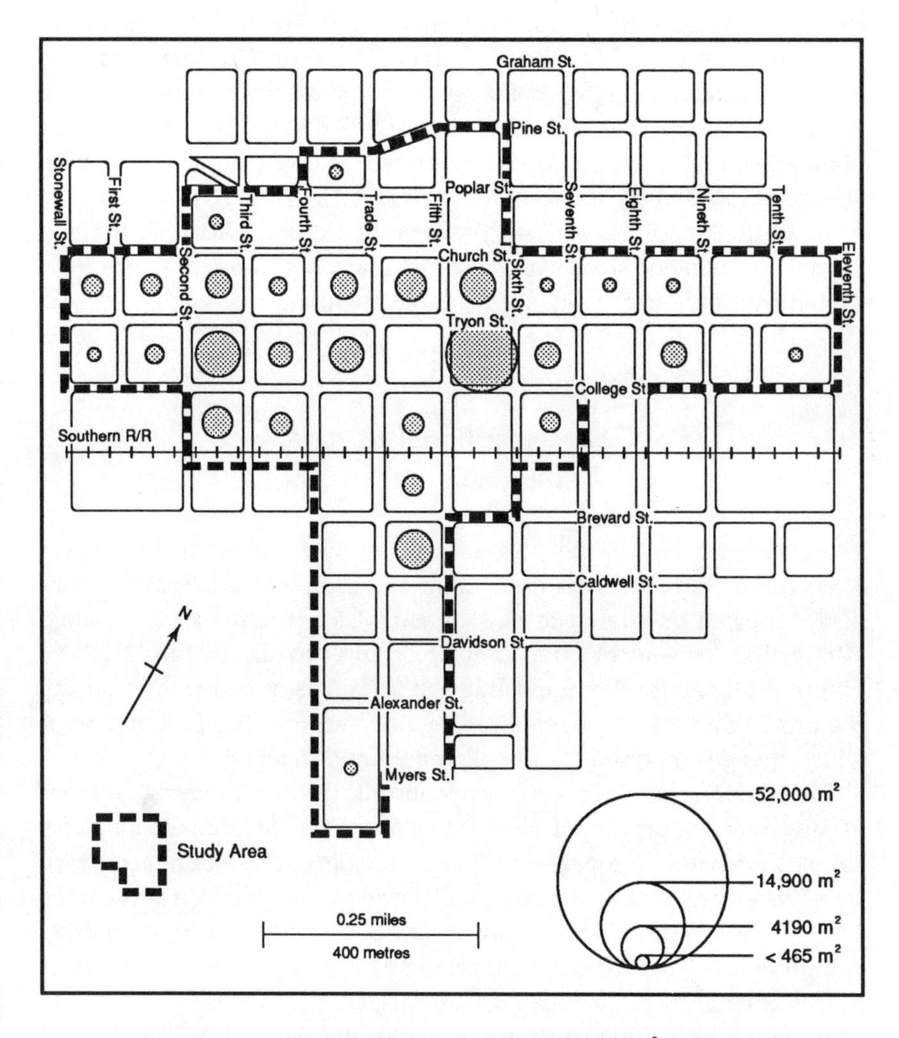

Figure 5.6 Charlotte city-centre retail floor area (m²), 1991.

Charlotte has experienced massive disinvestment in the central area. This is clearly shown by comparing the spatial distribution of retail floor area in 1991 (Fig. 5.6) with the position in 1972 (Fig. 5.3). The three large department stores, which had accounted for more than half of the total central-area retail space in the 1960s and 1970s, had closed by late 1990. Sears was the first to abandon downtown operations when it closed its store on the northeast end of the CBD in 1977, following the opening of a new store in Eastland Mall, a suburban regional shopping centre six miles east of the city centre. Belk followed in 1988 with the closure of its 37,200 m^2 (400,000 sq ft) store, now the site of the city's tallest office tower (the tallest office structure between Philadelphia and Houston). Iveys finally closed its doors in late 1990. The Sears store has been converted to an office building for use by county government, and the Iveys building was still vacant in 1992. Between 1970 and 1990, 74 planned shopping centres were built in the Charlotte area, most having suburban locations. These contained almost 930,000 m^2 (10 million sq ft) of retail floorspace, more than the total in Cardiff. Each of the three former department store companies in the central area is now represented by stores in the city's three large suburban malls.

Retail development and redevelopment in the 1970s and 1980s

The changes noted in the previous section have arisen in two ways. First, commercial premises are bought and sold, and leases are terminated, renewed or offered to new tenants. Thus premises may change into, within, or out of retail use, according broadly to the ability of different tenants to pay prevailing rents. This is of course affected by the profitability of various types of retail enterprise, itself a function of wider changes in systems of retail supply and demand.

The second type of change is more abrupt: the development, or more usually redevelopment, of substantial quantities of commercial floorspace. Property developers and financial institutions, when faced with a (re)development opportunity, will decide how much (if any) retail space should be provided. This decision in turn rests partly on market considerations – the demand for retail space, the rent-paying ability of potential tenants, and competing land uses.

In describing and explaining changes in central-area retailing of Cardiff and Charlotte attention focuses on the (re)development process.

Major changes in retail provision have occurred because of the commercial priorities of property developers and their financial backers. The centre of Cardiff has continued to be perceived as a major retail area by landlords, developers and investors. By contrast, the centre of Charlotte has become a major location for financial offices. Secondly, it is evident that in both centres retail success or failure can be ascribed partly to siting, layout and quality of the retail development that has taken place.

Cardiff

Figure 5.7 shows the current retail area and recent developments of

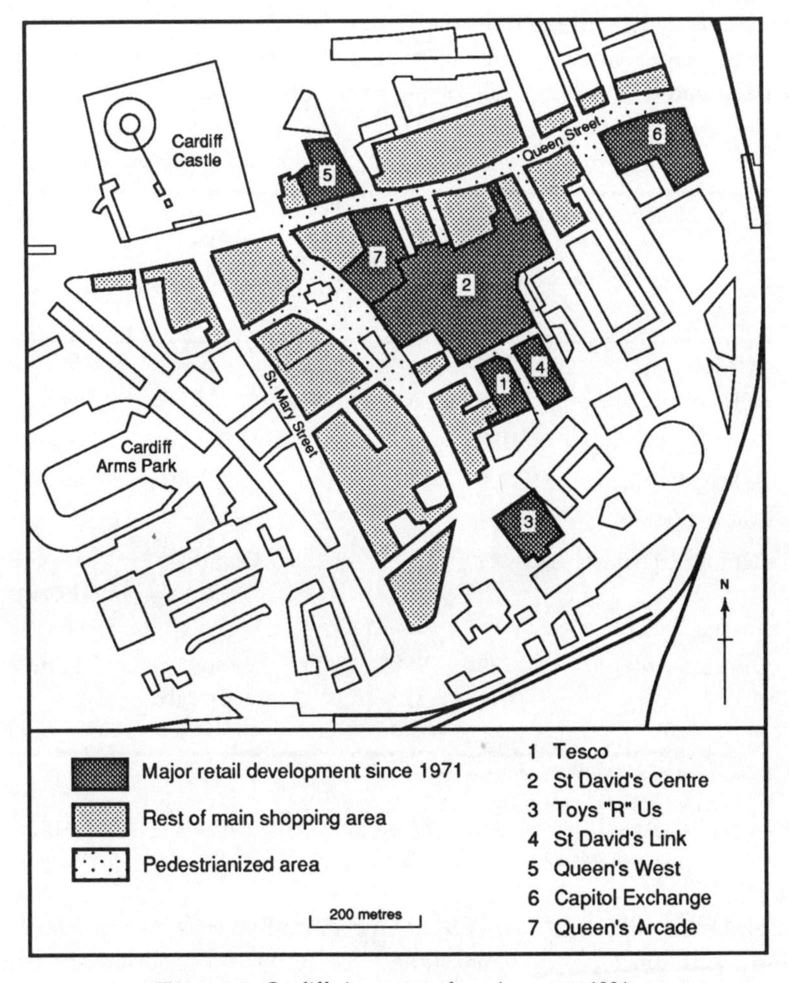

Figure 5.7 Cardiff city-centre shopping area, 1991.

retail space in Cardiff. When compared with the situation in the early 1970s shown in Figure 5.2, it is clear that the retail area is very similar in extent although the sites between the arms of the L are now partly filled in with new retail development.

Since 1980, the centre of Cardiff's retail area has been extended and modernized substantially by a series of shopping developments, the most important of which are listed in Table 5.3. These amount to about 83,600 m^2 (900,000 sq ft) of gross retail area (over 46,500 m^2 sales area) (500,000 sq ft). The first four schemes listed, completed between 1981 and 1987, were built in areas previously devoted mainly to non-retail uses. The three most recent schemes (1987–94) have been or will be built in areas where there was already a retail presence, and are thus redevelopments rather than totally new retail space.

It may seem strange that the gross retail area in the city centre rose by only about 13,900 m^2 (150,000 sq ft) in total between 1971 and 1990,

Table 5.3 Retail developments in Cardiff city centre, 1971–91.

Name of development	Year	Retail		Developer related to scheme	Other uses
		a	b		
Tesco store[c]	1980	1	4.2	Tesco	car park
St David's Centre	1981	66	41.8	Heron	none
Toys "R" Us store	1985	1	4.2	Toys "R" Us	ice rink[d]
St David's Link	1987	13	3.7	ARC	library
Queen's West	1987	22	5.9	Dixon's	food court
Capitol Exchange	1990	52	10.9	Guardian Royal Exchange	car park food court cinema
Queen's Arcade	1994	44	13.0	London & Edinburgh Trust	none?

a Number of shop units (at opening date).
b Gross retail floor area ('000 m^2).
[c] Vacated by Tesco in 1988; now occupied by a clothing store and a bookshop.
[d] On an adjacent site under different ownership.

despite the addition of nearly 93,000 m^2 (1 million sq ft) of new development. This discrepancy is explained partly by the replacement of old retail space by new development. More importantly, much of the older retail space in the city centre has been taken over by financial and leisure services, restaurants and takeaway food outlets. Unpublished survey data indicate that these service uses increased in area from around 27,900 m^2 (300,000 sq ft) to nearly 74,300 m^2 (800,000 sq ft) during the 1980s.

The first four developments shown in Table 5.3 are single-floor retailers, with storage and/or other uses on upper floors, except in the case of the Toys 'R' Us retail warehouse. The three more recent schemes (Queen's West, Capitol Exchange and Queen's Arcade) are all on two or three floors. This suggests a need for developers to intensify retail uses in order to justify development costs.

With one possible exception, discussed below, the new shopping developments have all been, or promise to be, commercially successful. This includes Capitol Exchange, opened at a time of serious recession in the retail industry, and yet largely let to reputable tenants (Table 5.4). The developers deliberately attempted to let this centre to up-market retailers, following a research study by the Management Horizons consultancy that identified a lack of high-quality multiple retailing in the centre of Cardiff. The older St David's Centre, devoted more to mass-market retailers, has been fully let almost from its date of opening, and has in the early 1990s reinforced its leading rôle through a £6.5 million refurbishment of its pedestrian circulation areas.

The least successful scheme commercially has been the Queen's West Centre, a redevelopment of the former British Home Stores variety store and some adjoining shops on the north side of Queen Street. This has never been fully let since its opening in 1987 (Table 5.4). This is probably because of its size and layout. Its 5,850 m^2 (63,000 sq ft) gross retail area has proved insufficient to accommodate an anchor store that would attract shoppers and other retailers. But its most serious problem is probably its layout: it has only one entrance from Queen Street, Cardiff's main pedestrian thoroughfare, and the other entrance is from a much less-frequented street so the centre does not lie along a heavily used pedestrian route. Several retailers have quit the centre, complaining of a lack of "passing trade". The upper floor units have proved particularly difficult to let: in 1989, some of them were subdivided into smaller units of between 22 and 42 m^2 (240 and 448 sq ft) in an attempt to appeal to independent retailers. This has not been entirely successful. There have also been closures among the more-accessible ground-floor units.

The national slump in retail performance of the late 1980s and early 1990s has clearly affected the viability of the Queen's West centre and the rate of take-up of units in Capitol Exchange. A further major effect has been the delay in commencement of the Queen's Arcade centre. The site became available with the closure of the Allders department store in 1986 and there were no major planning objections. The site was cleared during the following two years but building work was delayed until spring 1992. This delay appears to have been caused by lack of confidence in future levels of demand for expensive retail space, and possibly difficulty in securing the presence of an anchor retailer.

Table 5.4 Shop numbers and vacancies in new retail developments in Cardiff city centre, July 1992.

Name of development	Number of shops[a]		Vacancy rate
	Occupied	Vacant	(%)
Former Tesco store	2	0	0
St David's Centre	57	0	0
St David's Link	9	1	10
Queen's West[b]	17	12	41
Capitol Exchange[b]	40	8	17

Source: Field survey, 1992.

a. Total number of shops in each development may vary from that shown in Table 5.3, because of amalgamation or subdivision of shop units after the centre has opened. "Shops" include service units where located in the main shopping area of the development.

b. Excludes food court.

These recent problems have been repeated in other shopping centres across the UK during the past two years and there is no strong reason to believe that factors peculiar to Cardiff are at work. However, it may appear in retrospect that the 1980s were the "golden age" for retail development in the centre of Cardiff.

Charlotte

Real-estate development in Charlotte's central area was dominated during the 1970s and 1980s by the construction of several high-rise office towers; only a small amount of new retail space has been added. In late 1991, the central area contained 799,000 m^2 (8.6 million sq ft) of floorspace in multi-tenant office buildings, with another 102,200 m^2 (1 million sq ft) scheduled for completion in 1992. Some 85 per cent of this floorspace has been constructed since 1970. Much of the meagre new retail space has been constructed as part of either these large office towers or of new hotels, and so is not free-standing retail development. The only new free-standing retail scheme developed since 1970 is Cityfair, a festival centre opened in 1988 one block from the centre of the central area. This centre, built with substantial assistance from the city government, consisted of 5,390 m^2 (58,000 sq ft) of leasable space on three levels with an attached 678-space car park. Peak occupancy was reached shortly after its opening, but soon afterwards the number of tenants began to decline gradually. By the summer of 1991 only a few restaurants in the food-court area on the ground floor remained open. The centre closed in October 1991, but its owners announced plans to reopen it in autumn 1992.

Cityfair suffered firstly from the same difficulty faced by other central-area retailers for many years: the inability to attract residents from the city and suburbs to the city centre to shop. Second, and perhaps most important, was the decision of company officials to close the large Belk department store on the block adjacent to the Cityfair site. Plans had called for the Belk store to be attached to Cityfair by an overhead walkway. This walkway would have served to funnel shoppers between the two retail areas and would have provided convenient access to Cityfair for workers in several office towers via a system of overhead walkways. Cityfair was already under construction in 1987 when Belk decided to sell its property to a developer that had plans to redevelop the area with a 60-storey office tower and performing-arts centre. This mixed-use project was also designed to contain 4,650 m^2 (50,000 sq ft) of retail space. Shortly before Cityfair opened in October 1988, the Belk store closed. Thus, a major traffic generator and connector for Cityfair had disappeared. Further adding to the centre's woes was the construction of the mixed-use development itself: this created a barrier and isolated Cityfair from the large office workforce market in the downtown area.

Other factors contributing to the poor performance of Cityfair include its design and layout, and its failure to attract an anchor tenant. The main entrance at ground level was built facing away from the centre of the central area and the large potential market of office workers. Instead, it faced a low-income residential area that offered little market potential. In addition, the internal design of the centre, with the food court on the ground floor and retail units on upper floors, did not encourage "passing trade".

Local government support for the city centre

A final area of comparison between the two city centres lies in the degree of support for city-centre retailing provided by the respective local government administrations. Here there is some agreement in attitudes to retailing, but crucial differences in the legal powers available to the administrations.

Cardiff

In Cardiff, the city council and county council have consistently opposed major retailing development outside the city centre. They have used their powers under town and country planning legislation to refuse permission for such development on many occasions. While some of these decisions have been overruled on appeal to the Secretary of State for Wales (Guy 1988b, Lord & Guy 1991), there is no doubt that the councils' attitudes have encouraged retail development in the city centre. One commentator has stated that "Cardiff's retail market has been enormously encouraged by the city's refusal to allow development by the out-of-town brigade." (*Investors Chronicle*, 30 March 1990: 98).

The city council has also been involved directly in city-centre retail development. First, it has used its powers under town and country planning legislation to purchase land required for large-scale redevelopment schemes. In the late 1960s the city council and the Ravenseft property development company had agreed to carry out jointly a massive scheme of shops, offices and car parking, known as Centreplan, involving redevelopment of much of the central area. This project was abandoned in the mid-1970s property slump, when projected costs had already risen to £125 million without any certainty of sufficient levels of interest from prospective tenants. However, the city council had by then taken steps to buy several parts of the city centre from private landowners, using special powers granted by the Welsh Office. The largest of these areas, a site of some 3.2 ha (8 acres) to the south of Queen Street, was bought in 1977 by the council for £4.5 million. The council then entered into an agreement with the Heron Corporation (the developer) and Coal Industry Nominees (the financier) to build the St David's Centre on this site. The council is thus ground landlord for the centre, and also has a share in the centre's operating profits. The council also bought through compulsory purchase powers the site for the erstwhile Tesco scheme immediately to the south of the St David's

Centre (Fig. 5.7) and a large area at the southeastern tip of the city centre which now includes the Toys 'R' Us store, the adjacent ice rink and the Marriott Hotel (formerly Holiday Inn).

Secondly, the council has been involved indirectly in other recent schemes through detailed negotiation with developers. The most important and complex involvement has been in the Capitol Exchange centre. The original planning application for a shopping mall on this site was refused by the council in 1984, and this refusal was supported by the Welsh Office when the developer, the Guardian Royal Exchange insurance company, appealed against this decision. The refusal was partly for design reasons and partly for strategic reasons: the council felt that the retail element of the scheme (at that time, 18,400 m^2 (198,000 sq ft) gross retail, including a 9,300 m^2 (100,000 sq ft) department store), was too large and would adversely affect other parts of the city centre. Following the planning refusal, the developers and the council negotiated a revised scheme with a smaller retail content of 10,900 m^2 (117,000 sq ft) – some of the retailing was replaced by a multi-screen cinema – and more acceptable design features. In particular, the frontage of the scheme to Queen Street, the city's main shopping thoroughfare, was improved, so that retail units would open onto Queen Street as well as into the pedestrian mall within the scheme itself (Ralphs 1989). This revised scheme was approved by the council in 1986 and the scheme opened in late 1990.

Thirdly, the city and county councils have carried out an extensive programme of pedestrianization of the main shopping area. The whole of Queen Street has been pedestrianized since 1977 and further traffic-free streets have been added to form an extensive network that includes several covered "arcades" built mainly at the end of the 19th century. The councils have also provided public benches, tree- and shrub-planting, and signposting of pedestrian routes to major attractions in the city centre.

The county council has also been actively involved in city-centre retail development. It has been co-developer of the St David's Link development and has provided several multi-storey car parks around the edge of the core shopping area.

Charlotte

The decline in central-area retailing in Charlotte, already apparent during the 1960s and 1970s, did not result in any specific actions by city government and planning officials to stop or slow this trend until the 1980s. Three actions were taken during this latter period that were intended to strengthen retailing's presence in the central area. The first was the expenditure of $7 million in 1984 for a major facelift of the principal shopping street, Tryon. This included a widening of pedestrian sidewalks, provision of benches and bus transit shelters and tree-planting – all done along a 12-block stretch of the street. While this action improved the visual appearance, it had little effect on retailing along the Tryon Street corridor.

A second action taken by city government was aimed at preserving retail space at street level that was being removed by office-tower construction. City officials passed an ordinance in 1987 that required any new building with more than 9,300 m^2 (100,000 sq ft) of floorspace to devote 50 per cent of the net ground floor area to retail use. The potential of the ordinance to provide any significant amount of retail space was limited by the relatively small footprint area of even large high-rise office buildings. Furthermore, the limited space available meant that only a few small shops could be accommodated. Finally, the ordinance attempted to force the provision of retail space regardless of market conditions. Owners of two recently completed office towers have had difficulty in finding retail tenants for the street-level space.

The third action – the most direct and significant – was the involvement of city government in the construction of the Cityfair festival centre. The city spent $3.6 million to purchase this land for the centre in order to encourage the centre's construction by a private developer. It also loaned the developer $2.5 million from an urban development grant from the federal government and issued $6.1 million in bonds to build an adjoining 678-space multi-storey car park. Subsequently, the city became owner of Cityfair when the developer defaulted on the bank loan and the bank holding the mortgage wrote off its $18 million loan. In the process of looking for a buyer for Cityfair, the city refused to sell it without a guarantee that the building space would continue in retail use. A consultancy study commissioned by the city from the Urban Land Institute in Washington, DC, recommended that in the short term the city should invest money to make the space more leasable, but that it look for a developer in the longer term to build an office tower on the site. The centre was subsequently sold to an Oklahoma City firm that ran it until it was closed in October 1991. This company has announced that it plans to reopen the centre in autumn 1992 (coinciding with the opening of the adjacent 60-storey office tower), and to remerchandise the centre with off-price and factory outlet stores.

City officials and quasi-public organizations, such as the Charlotte Uptown Development Corporation and the Central Charlotte Association, have generally encouraged investment and economic development in the downtown area, but without giving priority to the preservation of retail space or enticing new retail development. These officials, organizations and local planning authorities have not shown the will, nor do they really have the legal means, to stop the conversion of central-area land from retail to other uses. Neither have they made any effort to stop suburban retail development, a position that would not be politically feasible because of the strong pro-development sentiment in the business and political communities.

Conclusions

This chapter has contrasted the histories of retail change in two similar-sized cities in the UK and US. From fairly similar positions at the beginning of the 1970s, Charlotte's centre has shrunk to almost complete insignificance in retail terms, whereas Cardiff's has grown in size and regional importance.

Commentaries on retail change in the two countries usually emphasize the effects of land-use planning regimes in explaining differences in retail growth patterns. It is held that both countries have undergone similar pressures towards suburban and out-of-town retail development, but that the British planning system has restrained such development, thus supporting the city centre. This explanation is clearly sustained by this case study. It is however by no means the full story. Contrasts in land-use planning control are reinforced by differing attitudes of financial institutions, shopping-centre developers, and multiple retailers to city centres. In the UK, city centres have been and are still regarded as the prime location for large-scale retail schemes, which can in normal times be funded entirely from private capital. In the US, city-centre retail development is seen as expensive and risky when compared with suburban or greenfield locations, and is unlikely to take place without substantial financial support from the public sector (Guy & Lord 1991).

Other, more local, factors have also been at work. First, the very rapid suburban growth of the Charlotte metropolitan region has on several occasions encouraged the development of major shopping facilities at what was at that time the edge of the built-up area. Population growth in Cardiff has been much slower. Secondly, the rise in importance of Charlotte as a financial centre, encouraged by deregulation of the US banking system in the 1980s, has led to pressure for major office development in the city centre. This is exemplified by the recent replacement of the Belk department store by a 60-storey office tower for NationsBank (formerly NCNB), which is now the fourth largest bank in the US. Cardiff has also experienced pressures for office development, but not for massive corporate headquarters on the Charlotte scale. These pressures have thus been diverted to town-centre fringe sites or, increasingly in the future, to the Cardiff Bay renewal scheme.

The case study also exemplifies the issues discussed in the introduction to this chapter. Retail decentralization has affected both cities, but in Cardiff has been restricted largely to convenience and household goods, leaving the city centre to concentrate on comparison goods and retail services. During the 1980s these were major growth areas in retailing and Cardiff city centre benefited as a result. In Charlotte, the department stores and mainstream comparison-goods stores have almost entirely vacated the city centre. Growth in these sectors is concentrated entirely in the suburbs. This typifies the differences between European and North American retail trends in recent years. Much of the recent uninformed and alarmist commentary on decentralization in western Europe has failed to understand this distinction.

The much-vaunted rôle of festival shopping is also called into question in this case study. In Cardiff, the city centre has remained the focus for investment, and festival shopping as such is not seen as necessary, although the recently completed Capitol Exchange centre with its food court and up-market outlets shows some influence of the festival shopping movement. In Charlotte, festival shopping has been attempted but has proved unsuccessful: neither the local consumer base nor the visitor/tourist base appear to be sufficiently strong to support it.

It would be hard to find a greater contrast in trends affecting the built environment in similar sized cities, than that described in this chapter. Yet there are some similarities between the two cities, not least the desire of local government councillors and officials to support retailing in the centre. The outcome has suggested that legal and economic imperatives are more powerful than political will in determining the characteristics of city centres.

PART THREE

PLANNING IMPLICATIONS OF RETAIL CHANGE

CHAPTER SIX
Prospects for the central business district

Gwyn Rowley

This chapter focuses on the central business districts (CBDs) of cities in the UK, their traditional importance in local space economies, and the considerable changes that are now occurring. These changes within the central-city retailing and service sector appear set to increase dramatically with the growing development of out-of-centre shopping complexes. Many British city centres now appear to be at a crossroads. The growth of further out-of-centre retail–commercial complexes could jeopardize the very basis of CBD existence and portend the end of the central city in its traditional form. These changes, and indeed the fundamental recompositions and selectivities within retail-business organizations, seem to be gaining in momentum. The effects of the large out-of-centre shopping centres (more than 46,950 m^2, 500,000 sq ft) on some CBDs are potentially calamitous, at least in the shorter and middle terms (Rowley 1989).

The problems and future prospects of the CBD are a major issue of contemporary retail change. The threat from out-of-centre development has been met by a continuing *laissez-faire* attitude from central government. Such out-of-centre retail developments as have occurred at Meadowhall, Sheffield, could undermine the very basis of CBD existence and its traditional regional dominance in the provision of comparison-shopping goods.

The inner city setting

There has been a general and continuing decline in many British inner-city areas, revealed in recent census analyses that show high concentrations of the unemployed, the lower-skilled or non-skilled, the elderly and ethnic minorities, accompanied by high levels of overcrowding, amenity-deficient housing and out-migration. Additionally suburbanization, counter-urbanization and "de-urbanization" tendencies have characteristically resulted in the higher socioeconomic groups decentralizing to the suburbs and beyond (Breheny 1990, Champion 1992).

Commentators on the inner city have generally ignored and largely overlooked the considerable importance of the CBD to the inner city; to its functions, especially retailing and commercial-service activities; and to related employments and space requirements (e.g. Cameron 1980, Lawless & Raban 1986, Balchin & Bull 1987, Lawless 1989, Robson 1988, Jacobs 1992). Such dynamic features of the CBD have been of quite considerable importance to the inner cities because of demand–cost relationships generated from within the CBD. Diamond (1991: 219) has recently gone so far as to suggest that: "The inner city . . . corresponds to [the] transition zone: it is an area that surrounds, but excludes, the city centre (or commercial 'downtown') and is itself surrounded by, but excludes the outer suburbs . . .". Such an arbitrary and simplistic distinction is categorically rejected in this chapter.

The viewpoint advanced here is that the CBD is an integral component of the overall inner-city space-economy. It is essential to recognize the rôle of the CBD in the inner city and also to retailing and commercial-service activities in general. The rôle of the CBD in the UK can be compared with its rôle in other countries. Although the focus is upon the British CBD, the evaluations must also be considered within a broader and more general spatio-temporal framework of inner-city changes that initially appeared in North America, are developing through Atlantic Europe and appear to be extending into Continental Europe (Hall 1985, 1988, Knee 1988, Rowley 1992).

The traditional CBD

In the 1960s Carol (1962) depicted what may now be termed the "traditional" position of the CBD: the primary business centre located at the commercial core of the city, offering the full range of retail-goods and service outlets, serving the entire city and its wider region (see also Murphy & Vance 1954, Carter & Rowley 1966, Ward 1966). Likewise Carter (1981: 198) viewed the CBD as "the organizing centre around which the rest of the city is structured".

A number of models were developed to describe the traditional CBD. Garner's (1966) concentric zonal model was based upon "general" accessibility and the retailer's desire to locate near the most central point of a concentration of commercial activities, given as the most trafficked spot. However, Nelson (1958) had previously noted that certain retailers, with "special" accessibility requirements, seek out adjacency to similar outlets by goods and/or quality, while others search for arterial accessibility with passing traffic. Davies (1972) subsequently presented a structural model of retail locations inside the core. This suggested that within the concentric arrangement of general functions, as posited by Garner (1966), it is also possible to discern clusters of specialized activities. These in turn would be overlain by a ribbon-component related to radial routes outwards from the CBD. Other work, on spatial associations within the CBD, emphasizes the linkages that help to maintain the overall attraction of the CBD for its entire regional population (Getis & Getis 1968, Davies 1973, Shepherd & Rowley 1978, Brown 1987).

The changes under way both within and beyond the city centre (Davies & Sparks 1988) may necessitate an abandonment or at least a qualification, reformulation or extension of these "traditional" monocentric approaches. The approaches of Garner & Davies now have to be reconsidered in the light of the changing nature of retail locations, which derive from changing accessibility and the attractiveness of alternative commercial locations (Rowley 1990). This will require a far more explicit attention to independencies, as shopper access at larger, more convenient, possibly out-of-centre locations increases (Lord 1988). Indeed, the economic advantages of the CBD have usually been considered under the catch-all concept of agglomeration economies, but these might be interpreted more appropriately as economies of centralization. From this it has been suggested that: "any off-center subcenter would do just as well, provided it is reasonably accessible" (Richardson 1978: 48). Thus it is possible to argue that a ring of subcentres might satisfy shopper preferences even more than a single centre, and any existing force of agglomeration economies will not necessarily save the CBD.

Some elements of the models and theories referred to above continue to have relevance. Nelson's (1958) ideas on the "special" accessibility requirements of certain retailers either under "adjacency" considerations and/or "arterial access" may enable us to understand some of the changes in business locations now under way. Also, Davies's structural model of retail locations inside the core, in identifying three components – nucleated, ribbon and specialized – can also provide a basis for such an extension. It is perhaps the ribbon and specialized components of existing and potential CBD businesses that may initially be particularly susceptible to the attractions afforded by out-of-centre locations.

The changing CBD

While the absolute peak of the CBD may remain, local commercial maxima can come into existence in certain subcentres. The continuing dominance of the central retail nucleation has to be viewed alongside the fundamental changes that are occurring in British retailing and, in particular, the continuing upsurges at out-of-centre locations (Rowley 1986, Dawson et al. 1988).

The initial advantages of the CBD, in essence, derived from the sum of the agglomeration and transportation advantages, as the best location for serving the population of the entire city and wider region as a whole. These advantages may not continue indefinitely, for with city growth and spatial expansion, transport (congestion) costs increase and at some level of scale (city size) the agglomeration economies will be weakened. Here a stochastic element in locational choice will arise where the probability of a new locator or a potential relocator choosing a CBD site will be a function of the CBD's perceived net benefits, as opposed to alternative available locations. In the longer term, similar choices will confront an existing operator in maintaining a CBD location.

Over time, the failure of the CBD to attract and retain certain outlets and their agglomeration at other sites tends to become cumulative. Perceptions of relative advantage and movement may be based on decisions by "key leaders". For example, if certain "critical" stores are breaking rank in their particular association with the CBD, others might simply follow. Later, it will be shown how certain central shopping districts are experiencing reductions in, or failing to keep pace with changes in, the provision of certain types of goods. If these goods figure prominently in overall trip motivations or comparison-shopping circuits, such centres might be by-passed, with consumers proceeding to a centre where a larger range of goods is on offer.

In North America retail growth centres are now generally at out-of-centre locations (Dawson 1983, Lord 1988, National Audit Office 1990). Whereas before 1920 in the USA about 90 per cent of total retail sales in urban areas were from downtown outlets, they are generally below 50 per cent in most cities and are still declining (Yeates & Garner 1980: 336). Garner cited the specific example of Portland, Maine, where by

1972, accompanied by the development of 10 outlying shopping centres in Greater Portland, CBD retail sales had slumped to 28.6 per cent of the 1958 level (Yeates & Garner 1980: 336). Portland Chamber of Commerce data show that by 1991 Portland CBD retail sales had declined even further, to less than 11 per cent of the 1958 level.

Thus what was viewed as the essential ordering mechanism in the Garner model, relating to the interplay between principles of general accessibility to the centroid and the locations of the various orders (thresholds) of goods and services, appears to be in need of major reassessment and amendment. Such a refinement will require reference to the fundamental changes in the overall attractiveness ("access") of the CBD to the entire city for the complete range of goods and services and particular specific components of that range.

Davies (1972) presented an interim model of the changes that appeared to be under way in certain British CBDs. Some of the special area characteristics Davies identified, such as entertainments, furniture and appliances, have recently grown up at out-of-centre locations, usually with a related decline in CBD provision of these "goods". Table 6.1 suggests types of overall recent change. The key question is whether such declines within the centre will spread and deepen. Will further gaps in retail provisions appear and are the levels of depletions and vacancies in our CBDs set to increase noticeably?

Table 6.1 Characteristic functional changes in/around British CBDs, 1987–92.

Function	CBD	CBD interceptor	Fringe CBD	Out-of-centre shopping centre (other out-of-centre)	Rural small town
Apparel	–	+	*	++	+
Variety	–	*	*	+	*
Gifts	–	+	*	++	+
Food/groceries	–	*	*	+(+)	*
Banking/financial	*	*	*	+(+)	*
Cafés	–	*	*	+	*
Garages (car showrooms)	–	*	+	(+)	*
Entertainment (e.g. cinemas)	–	*	*	+	*
Market/mart	–	*	*	(+)	+
Furniture	–	*	++	+(+)	*
Appliances	–	*	+	+(+)	*

– decline   * stable   + increase

New growth centres and the CBD

As alluded to above and elsewhere (Rowley 1985, 1986, Davies & Sparks 1988, Jones & Simmons 1990) established CBDs have not generally been attracting outlets and sales in significant growth sectors of retailing, such as furnishings, electrical goods, garden centres and DIY and are in fact losing their share. When set in the context of changes in patterns of consumption-expenditure, it is seen that the types of retailing developing at out-of-centre locations often include many of those "more-profitable" ranges and innovatory forms. Indeed the scale developments at out-of-centre locations also proceed in tandem with evolving marketing practices. For example, the progressive move over to larger out-of-centre stores by the large multiples links with strategies to entice customers to purchase more fancy "exciting" products: such as replacing or augmenting purchases of baked beans with fresh, stuffed pasta, or with different versions of baked beans at higher margins, for example, low-sugar, low-salt "healthy" baked beans. Such stores are also particularly good at selling "added-value" products like ready-made meals for which customers pay a premium for attractive packaging and convenience. The larger stores, by attracting customers to move to these "fancy" products and by spreading costs more thinly, are much more profitable, even though the actual prices of comparable products are the same. The larger multiples, Sainsburys, Tesco, Asda and Safeway, have recently embarked on vigorous superstore programmes. Various data indicate that among these larger multiples selling areas have more than doubled since 1980, as have sales per square foot, and sales per store have quadrupled or quintupled. Of course, as indicated above, these superstores (over 2350 m^2, 25,000 sq ft), replete with adjacent large car parks, are generally at out-of-centre locations.

One of the traditional CBD strengths is its overall attraction to consumers as an ideal centre for comparison shopping, where the assemblage of ranges of items in proximity to each other facilitates cross-shopping. However, even this CBD advantage is threatened. It is precisely these areas of significant cross-shopping, for example apparel and luxury items such as jewellery and leisure-sporting goods, for which the large out-of-centre centres are now competing. If expansion in such retail sectors were to gain in momentum at out-of-centre locations the competition might affect the underlying vitality of certain established CBDs, where for example over 40 per cent of their comparison-goods outlets are related to the apparel sector.

It should also be recognized that throughout the 1980s about 10–12 per cent of stores within a typical CBD were changing hands each year. Evaluations being made by Rowley indicate that during the 1990s the turnover has been far faster and that downward trends in the established balance between store closures and openings are increasingly apparent. As long-established traders pull out, more and more sites are either left empty or else taken up by down-market short-lease discounters, paying low rents to desperate landlords. Furthermore, city-centre rents are relatively unresponsive to falling demand: most leases are signed for fixed rents subject to five-year upwards-only reviews; so rents can fall only when leases change hands, not when existing tenants renegotiate their five-yearly rents. This means that tenants with leases fixed in 1988–9 may have been hit by high costs as well as by low consumer spending.

The trends in the British CBD associated with the emergence of new growth centres can be seen to relate to Meier's (1985) report on the broader range of retailing and urban services in California. There, new agglomerations of activity were emerging as "new metro centers . . . competitive with the central business district", where "a more diverse stock is now maintained outside the downtown area" (Meier 1985: 75). This contrasts markedly with the "traditional" position of the CBD, considered by such as Carol (1962) and Carter (1981), alluded to earlier.

The continuing work of Goad (1987) and Hillier Parker (1991) points to the dimensions and ebullience of many out-of-centre developments in the UK (see also Gayler 1989a, Jones 1991, Hallsworth 1991b, Williams 1992). For example, by late 1992 five major out-of-centre regional shopping centres of over 46,950 m^2 (500,000 sq ft) had been opened, while another 38 or more were at various stages of planning and development (Fig. 6.1). Although Schiller suggests (personal communication, 1989) that only 12 to 15 of these centres will become regional shopping centres, many of them will aggravate the changes already occurring in the CBD.

Another disturbing trend in the British context concerns the commercial–professional office functions. These have for long been associated with the city centre and their strength has been considered likely to limit decline in the CBD. Nowadays, however, business parks are attracting not only computing functions, a traditionally footloose category, but also increasingly banking, insurance and financial services, to compete with town centres (Reid 1991). Indeed there may be an analogy here with retailing: retail parks were first occupied by users that did not compete directly with town centres (DIY, bulk furniture and discount

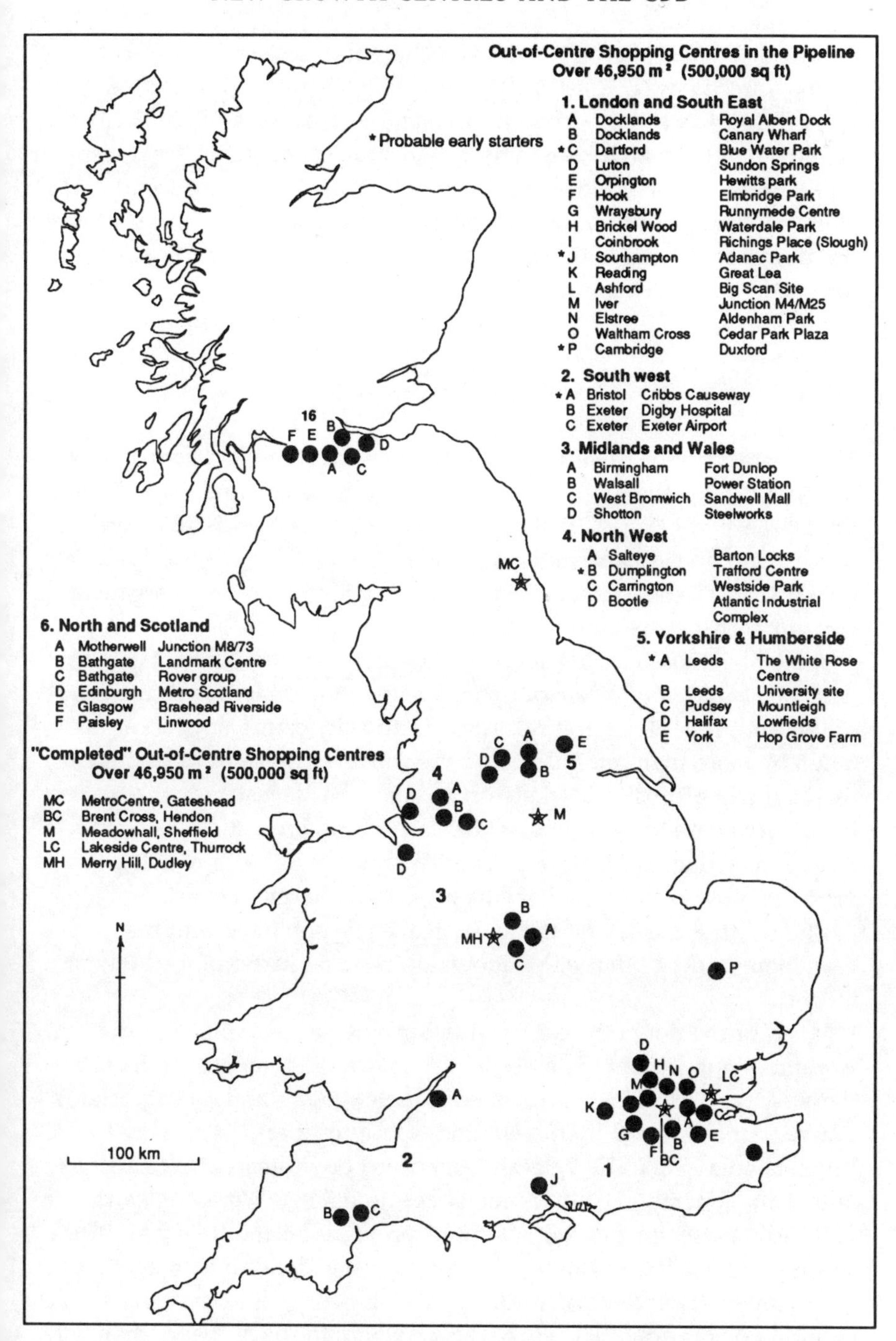

Figure 6.1 Major out-of-centre UK shopping centres in the pipeline, 1992.

electricals). Subsequently, however, the parks acquired retailers competing directly with those in the CBD (e.g. Toys 'R' Us and Next at Fosse Park, Leicester). Computer-related companies and similar users are now being joined increasingly by the more conventional town-centre office users on the larger business parks. This developing trend looks likely to erode further the overall commercial–office primacy of the traditional town centre in the UK. In turn, this will reduce the demand for goods and services provided for office workers in the CBD.

Decay in Sheffield CBD

The situation in the Sheffield CBD illustrates the trend of decay. In order to examine the changes, Sheffield's CBD can be conveniently divided into three parts (Fig. 6.2). Haymarket/Castle Markets is the northern section running down slope from High Street, along the Haymarket towards the Castle Markets. The central section focuses on Fargate and the southern part is around the Moor.

Since the massive Meadowhall complex of more than 116,125 m^2 (1,250,000 sq ft) of retail space opened some three miles to the northeast of the CBD in September 1990, trade in the city centre appears to have fallen by more than one-third, and the entire northern third of the CBD, the Haymarket/Castle Markets area, has been devastated. Streets are lined with boarded-up shops, whitewashed windows and down-market discounters (Fig. 6.3). Several charity shops now occupy what were prime retail sites until the later 1980s. C & A, the only department store in this northern part of the CBD that does not have another outlet elsewhere in the CBD or at Meadowhall, shed a quarter of its floorspace in 1991.

Much of the down-market retailing in this northern third of the CBD is rented out on short three-month leases and ranges from cheap footwear and clothing to tatty household goods and charity shops. Overall, the ground-floor retailing space of the Haymarket/Castle Markets area is one-third vacant, one-third down-market retailing and one-third general, mostly convenience, retailing. Were BHS and the House of Fraser to pull out, it could prove to be the final nail in the coffin. This area is undoubtedly experiencing decay and many locals increasingly steer clear of it during the day as well as the night. The confidence of potential developers seems to have been seriously undermined and more favourable investment opportunities are seen to exist elsewhere.

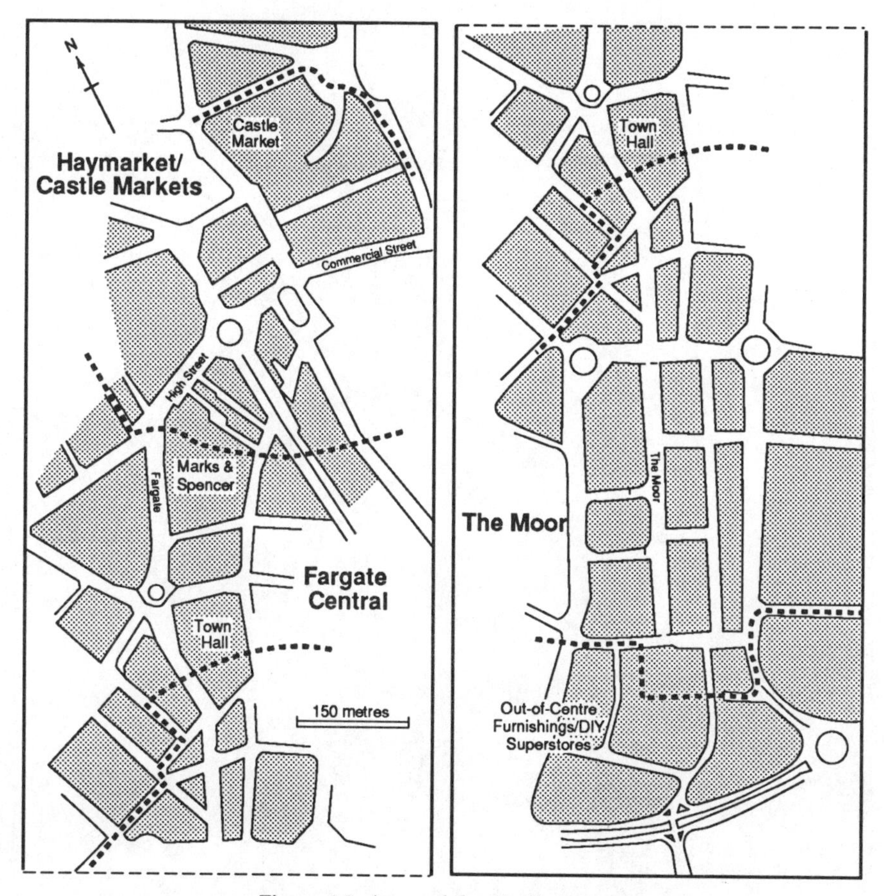

Figure 6.2 Areas of the Sheffield CBD.

While the central section of the CBD around Fargate retains a certain ebullience there are signs of decay there as well. The Marks & Spencer store is running a definite second to the company's Meadowhall branch and ranges of goods are consequently relatively limited. The local John Lewis store, Cole Brothers, still survived in 1992 but a number of smaller retailers, for example the Western Jean Company, a fashion store, had closed down and moved to Meadowhall. Lingering vacancies and downgradings are increasingly evident on the periphery of Fargate, as in Orchard Square to the southeast and Pinstone Street to the south of this area.

The southern section of the CBD, around the Moor, is seeking a specific market niche in what can be termed a bazaar economy with cheap street-market stalls along the pedestrianized Moor, which also offers brass band concerts and dancing majorettes. While such energetic

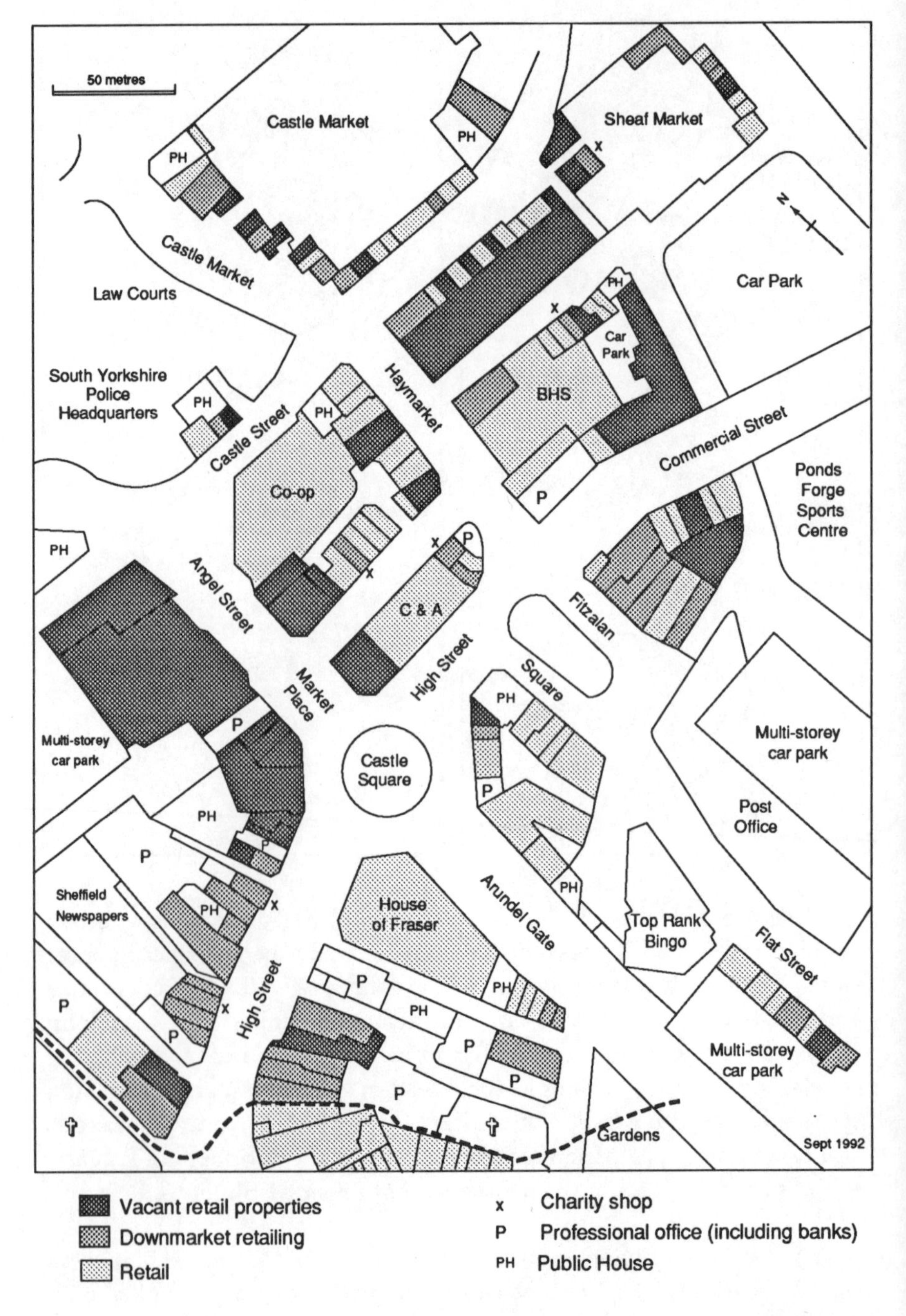

Figure 6.3 Retail property in the Haymarket – Castle Markets area of the Sheffield CBD: September 1992.

activity is welcome, the general air is of a non-CBD local shopping district that seems set to survive at a reduced threshold-capacity level. Marks & Spencer closed its store on the Moor when it opened its Meadowhall branch.

Overall the Sheffield CBD is seen to be in major decline, with confidence falling and little likelihood of significant private investment in the near future. Of course without re-investment, further decline will occur and the downward spiral could accelerate.

General trends in the British CBD

The evidence from other British cities is conflicting. Like Sheffield, central Dudley has been particularly hard hit by the huge Merry Hill shopping complex, with Marks & Spencer, British Home Stores, Littlewoods, Sainsburys, Next, Tesco and C & A all leaving the centre. Further evidence on the decline in Dudley is presented in Chapter 7. Newcastle upon Tyne, however, appears to be avoiding the trend, despite the 1986 opening of the MetroCentre across the river in the Gateshead Enterprise Zone. Large-scale and early redevelopment and promotion of the city-centre buildings around Eldon Square, combined with the city transport improvements of the 1970s, have enabled the Newcastle CBD to retain its thriving retail function. Moreover, Newcastle has further peculiar advantages: first, it is the only large city between Leeds and Edinburgh; secondly, the northeast has not been hit to the same extent as other parts of the UK by recession of the early 1990s. Not many cities can expect to follow its example.

The planning response

Fundamental changes, reorganizations and developments are under way in British retailing. Developments at out-of-centre locations have threatened the traditional hierarchical structure and the CBD. The emergent huge out-of-centre retail developments of more than 46,950 m^2 (500,000 sq ft), with three–six anchor stores and at least 100 other store units, are the largest of the new developments. No less dramatic are the suburban retail parks, where a couple of superstores with free car parking, selling bulky goods, such as B & Q or Asda, have increased from one in 1980 to more than 250 in 1992.

A significant take-off in the development of these large out-of-centre retail complexes is likely unless they are constrained by a central government planning response. Two background matters of particular and mounting concern need to be considered: the ambiguity of central government policy and reductions in the powers and discretion of an increasingly emaciated local government (Rowley 1984, Audit Commission 1989, Home 1991). The creation of enterprise zones and urban development corporations, for example, has removed significant segments of urban regions from local authority control. Many of these areas are avidly sought out by the larger retail-commercial developers (Rowley 1986).

The current Planning Policy Guidance 6 (PPG6) on major retail development (DOE 1988, 1992b) fails to present a clearly defined policy. PPG6 (DOE 1988) essentially represents a consolidation and restatement of earlier pronouncements. NEDO (1988a) subsequently called for a clear, coherent planning framework and a far firmer set of guidelines, although it eschewed a blanket prohibition on large out-of-centre developments. However the recent examples of developers gaining ministerial consent for massive out-of-centre retail complexes, after having successfully appealed against earlier planning rejections at both local and national levels, serves to illustrate the ambiguity. In 1987, for example, the Secretary of State for the Environment refused planning permission for the proposed Cribbs Causeway development in view of "the damage which could be done to the vitality and viability of the city centre of Bristol as a result of the redevelopment of [the] Broadmead [centre] being held back" (Howard 1990: 166). This accords with PPG6, which refers to developments "on such a scale and of a kind that they would seriously damage the vitality and viability of the town centre as a whole" (DOE 1988).

The ministerial refusal of planning permission for Cribbs Causeway was in fact cited by Michael Howard, the Minister for Water and Planning, as an example of a situation "where the threat to the town centre is legitimately removed" (Howard 1990: 166). However, in November 1991, Michael Heseltine, the Secretary of State for the Environment, granted planning permission for the development of Cribbs Causeway. This is a Jack Bayliss–Prudential scheme with "outline consent" at about 61,000 m^2 (650,000 sq ft), but with adjacent land that will finally bring it up to more than 93,900 m^2 (1 million sq ft), set beside the M5 motorway north of Bristol, near the M5–M4 intersection. Its estimated cost is well over £150 million and it is planned to have more than 150 stores. Its prime interceptor location for regional trade from the South West, South Wales and the South Midlands will probably assure its success. Significantly, Cribbs Causeway lies outside Bristol in North Avon. The Broadmead centre in the Bristol CBD is unlikely to retain its competitiveness status in view of the long-standing hostility of many Bristolians to its soulless character: it is an area of the CBD rebuilt following extensive Second World War bomb damage. Its placelessness, lack of parking and also recent problems of safety and adjacent inner-city disturbances, could all contribute to a potential devastation of the Bristol CBD.

A similar out-of-centre development under consideration is the White Rose Centre, Leeds, a scheme of more than 56,340 m^2 (600,000 sq ft) set some four miles out from the CBD on the southeast fringe of the city. Again, following appeals, permission was granted by the Secretary of State for the Environment in 1988. A start here will probably have to await the end of the recession and lower interest rates. In the meantime, in 1991 Leeds appointed a town-centre manager charged with improving the image of the CBD, through such measures as the reduction of graffiti and litter, the improvement of street furniture and tree-plantings. However the eventual opening of the White Rose Centre is still likely to be ominous for the Leeds CBD.

The prospects for the Manchester CBD also look increasingly bleak. This CBD could be in a precarious state by the mid-1990s when the development of Trafford Park, set in industrial wasteland alongside the Manchester Ship Canal, gets under way. This particular case reveals a fundamental dynamic of the entire process: the probability that property developers can realize quick and massive profits in the development of derelict lands in collusion with increasingly desperate local government administrators.

It seems that councils are now being intimidated into granting planning permission for larger retail developments by the threat that they will have to pay heavy compensation to developers if they attempt to withhold consent (Broom 1992). In particular, Broom reports, councils that refuse plans on "environmental grounds" are vulnerable to cost orders on the basis that "protecting the countryside" is not a valid reason for refusing a planning application. The combination of weakened local governments and unclear national guidelines could make it a field-day for would-be developers (cf. Knee 1988).

The strength of present trends is demonstrated by a public enquiry held in 1991 on the development of a large (more than 46,950 m^2, 500,000 sq ft) retail complex, that was included in the Cambridgeshire County Council Structure Plan of 1989. The decision of the DOE is expected in March–April 1993. It seems probable that the development will be allowed and that of the four possible sites isolated that at Duxford will be accepted. The Duxford development by Grosvenor Developments–Tesco would be near the M11, midway between Cambridge and Saffron Walden. The John Lewis Partnership has intimated that it will move to Duxford from its cramped store in central Cambridge. Whereas this development would sap some of the retail-commercial vitality of central Cambridge it has sufficient other strengths derived from its university–tourism markets to survive. Yet the Cambridge case is indicative of the spate of outward developments that could develop around even smaller British towns and cities.

Private transport access to the established CBD is another matter of concern. Plans to limit drastically vehicular access to British city centres are now gaining broader acceptance. The European Community scheme "Cities without cars" unveiled by the then EC Environment Commissioner Ripa di Meana in August 1992, the Traffic Calming Act of March 1992 and possible road-pricing schemes to reduce central area traffic are likely to constrain the use of cars within city centres. However this might simply induce a further significant downturn in CBD trade, particularly as the more affluent may wish to continue to use their cars for shopping (Kern et al. 1984). The wealthier are likely to switch their patronage to the new out-of-centre developments. The poorer, less affluent and less mobile inner-city residents would be left as customers of the declining CBDs. The cases of Hartford, Connecticut, and Buffalo, New York, and of numerous other US downtown areas have demonstrated this trend.

It is possible that trends in the UK will follow the US pattern of continued inner-city and CBD decline that in 10 to 15 years will be followed by US-style attempts at revitalization programmes (DRDC 1988, Frieden & Sagalyn 1989, Feinstein 1991). Cullingworth (1989) points to the need to consider the experience of Canada, where downtown areas have not declined in the same way as in the USA. National, provincial and city governments within Canada, it must be said, have appeared far more able and ready to orchestrate initiatives to maintain their CBDs than their present British counterparts (Rowley 1984).

Concluding comments

The new large regional shopping centres and other out-of-centre developments could have a quite dramatic and possibly catastrophic effect on the commercial futures of a number of established British CBDs. Of course a certain ebullience might continue in some of the more favourably situated or already redeveloped centres, although it is likely that they will lose their traditional overall primacy within the regional retail hierarchy.

This chapter has focused attention on the CBD of British cities. It is emphasized that "traditional" ideas of CBD dominance and structure now require serious qualification and reformulation. The CBDs may be seen to be "missing out" on significant growth sectors within the broader retail environment and the continued monitoring of functional changes and consumer behaviour assumes vital importance. There may soon be an increased movement of certain stores, traditionally associated with the CBD, to those developing out-of-centre constellations, and a downgrading and in time even a progressive abandonment of their outlets within the CBD. There is now a pressing need for improved data on UK retailing, particularly on geographically referenced retail turnover, to facilitate a fuller and deeper consideration of retail structures and change (Guy 1992).

Back in the late 1970s, studies of US cities by the Brookings Institute, which focused upon high unemployment rates, deepening social polarizations, urban land "severely under-utilized with inner city land dereliction resulting not only from decentralization but also individual mobility", increasingly came to the view that the problems of these old, declining and isolated core cities represented "the domestic problem of the United States" (Nathan & Dommel 1977: 10). The deepening problems of the CBD and the inner cities in the UK now demand a move beyond the short-term to longer-term strategies. Otherwise the inner cities and their CBDs will continue to decline and deteriorate. These are indeed critical issues that lead on to problems of the overall importance of the CBD within our larger cities. Likewise, the CBD must be seen as a special feature of the inner city, a bonding mechanism for the wider urban region, and as an important component in the lives of inner-city residents. It is time that effective planning controls were introduced to save the British CBD, yet this now appears to be a highly unlikely development.

CHAPTER SEVEN
The impact of out-of-centre retailing

Colin J. Thomas & Rosemary D. F. Bromley

The emergence of out-of-centre retailing has transformed the retail structure of most cities in western Europe and North America (see Chapter 1). The new facilities have undoubtedly had a major economic, social and environmental impact on the urban scene. Its varied character and the response of planners are the key issues considered in this chapter. Attention is focused particularly on the changing patterns of shopping behaviour indicated both by studies of the impact of the new facilities and by new evidence drawn from a number of large-scale household and city-centre shopper surveys in the UK. Changing consumer behaviour, expressed in the transfer of trade from the older shopping facilities to the new, has critical implications for the vitality of the traditional retail system. Existing retail investment is under threat and the less mobile elements of the population are experiencing diminishing access to the newer shopping facilities. The commercial and planning implications of the range of evidence reviewed are also discussed.

The impact of superstores and hypermarkets

Early studies of the impact of superstores and hypermarkets on patterns of shopping behaviour focused on the trading characteristics of individual stores, and regular patterns emerged (e.g. Thorpe et al. 1972, Thomas 1977, Lee Donaldson Associates 1979). Most visits (usually greater than 60 per cent) were weekly bulk grocery shopping trips for overwhelmingly car-borne shoppers drawn from the middle and upper status groups. The stores also had particular appeal for the larger family groups associated with the young family-forming and middle stable-family stages of the life cycle. High degrees of grocery shopping allegiance were demonstrated, with typically about 80 per cent of shoppers purchasing more than 75 per cent of their goods at a chosen store. The combined advantages of convenience, comprising ease of access and parking, long trading hours and the opportunity for bulk-purchasing; price competitiveness; and a good quality and range of products, were potent attractions for these market segments. Lower status groups were under-represented since they lacked the mobility and the ability to make the large financial outlay associated with bulk-shopping trips, while long-established shopping patterns, mobility constraints and smaller family sizes appeared to deter the older age-groups. Trade areas, while displaying a distinct distance–decay effect, were far more diffused than traditional shopping centres of similar size, and car-borne shoppers were drawn in steadily increasing numbers from distances of up to 10 miles. Characteristically, upwards of 10 per cent of customers travelled from even farther afield.

Despite initial fears about the adverse impact of these developments on town centres, the evidence suggests that their effects were felt most by the smaller district centres in their immediate vicinity rather than by the central business districts (e.g. Thorpe et al. 1972, Thomas 1977). More recent evidence continues to support this view. The opening of the Tesco hypermarket at Neasden in 1985, for example, resulted in a 11 per cent loss in customers and a 26 per cent loss of turnover at the nearby Blackbird Hill supermarket. Almost all of this loss was accounted for by car-borne shoppers who transferred to the new store (London Borough of Brent 1986). A more extensive longitudinal investigation in three stages of the Carrefour hypermarket at Caerphilly is also worthy of note (Lee Donaldson Associates 1979). Over the first five years of trading a reduction of 11 per cent of foodstores was detected in the trade area, and the nearby Caerphilly town centre (district status) appeared to be most adversely affected. However, a widespread negative impact was not demonstrated and Caerphilly continues to be a viable local shopping centre. In common with most of the superstore and hypermarket developments their appeal to particular consumer sub-groups drawn from extensive trade areas results in the dispersal of their impact widely across the traditional hierarchy.

Thus, since the late 1970s initial fears about the potentially drastic effect of superstores on older shopping facilities has gradually declined and superstore trading has emerged widely as the major feature of grocery shopping for the more mobile sections of the community. In fact, the superstore now forms the cornerstone of the operations of all the major grocery retailers in the UK and most large cities have accumulated a number of superstores in strategic out-of-centre locations in each "sector" of the city.

Evidence from Swansea

The Swansea District is typical of this pattern: between the late-1970s and mid-1980s six superstores were developed, two of which traded in close proximity to each other in the Swansea Enterprise Zone (Fig. 7.1). A large-scale household shopping survey undertaken in 1986 throughout the District (1991 population 182,100) offered the opportunity to assess the cumulative impact of a number of superstores on shopping habits in a British city (Bromley & Thomas 1990).

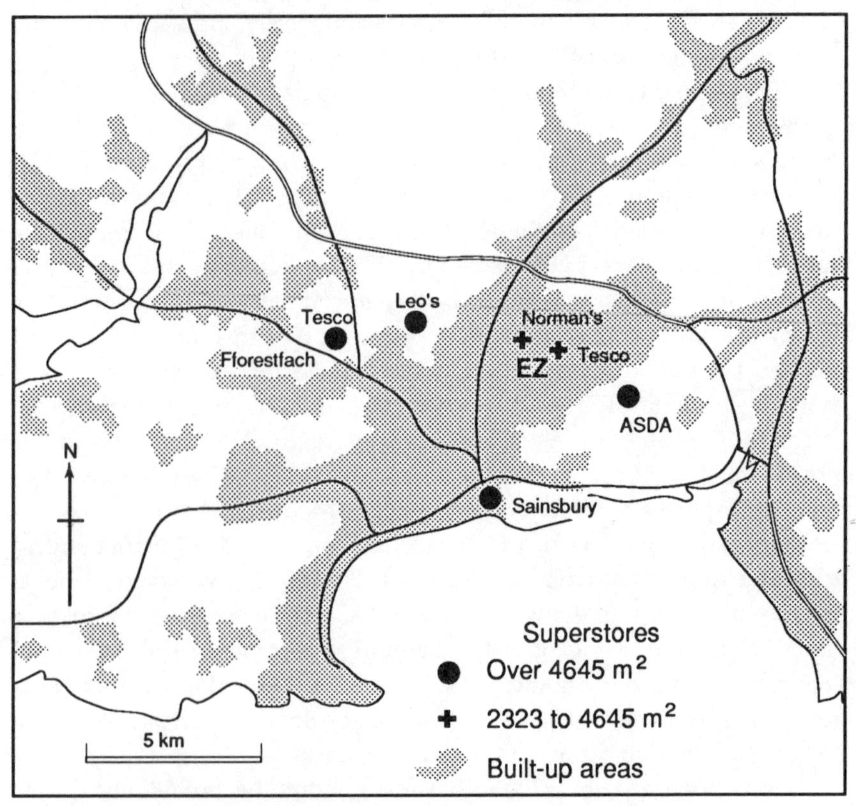

Figure 7.1 Superstores built in the Swansea area, 1977–86.

A 3 per cent random sample of households from the 23 wards of the city, consisting of 2,199 respondents, was grouped into six broader residential sectors for analysis. Each sector exhibited a reasonable degree of social homogeneity and provided sub-samples ranging from 204 respondents in Gower to 590 for the central wards (Table 7.1). The southwestern and western sectors house a middle to upper status suburban population, while the Gower sample is of similar social type but lives in a more dispersed rural situation. However, the northern, eastern and central sectors tend to house somewhat lower status groups living in small terraced dwellings or local authority housing estates.

Consistent with the behavioural norms of an increasingly car-oriented society, the destination of the "main" food shopping trips displayed considerable spatial variability (Table 7.1). No single destination dominated the grocery shopping trips of the residents of any of the six sectors. Nevertheless, on average, the newly developed superstores together attracted most shoppers (58 per cent). In fact, for the areas that offered a choice of superstores, such as the northern and eastern sectors, the superstores accounted for more than 70 per cent of main grocery shopping trips, despite being the areas with the highest concentration of households that did not own a car (40 per cent and 36 per cent, respectively).

Table 7.1 Location of main food shopping in Swansea (%).

Area*	n	City	Districts	Sainsbury	Tesco Ffach.	EZ s'store	Leos	Asda	Local
Gower	196	20.4	8.7	25.0	32.7	1.0	0.0	4.1	8.2
South-west	214	15.9	41.1	29.0	7.9	2.3	0.9	0.9	1.9
West	416	19.5	8.4	12.3	40.6	1.7	6.0	2.6	8.9
North	445	9.7	13.9	4.5	14.4	30.3	13.5	7.6	6.1
East	291	18.6	3.4	12.7	0.7	35.4	0.0	27.1	2.1
Centre	568	38.2	11.1	9.7	14.6	5.6	4.8	4.2	11.8
TOTAL	2130	22.0	12.9	12.9	18.7	13.3	5.4	7.4	7.4

Source: Household survey, 1986.
* See Figure 7.2.
Key: EZ = Enterprise Zone Ffach. = Fforestfach

By contrast, the proximity and attractions of the city centre resulted in the lowest level of superstore patronage by the residents of the inner-city wards (39 per cent) and the continued strength of the city centre for these respondents (38 per cent). The low level of superstore patronage by the mobile higher status residents of the southwestern suburbs (41 per cent) is explained by the absence of a superstore in this sector of the city, combined with the presence of a strong district shopping centre (41 per cent). Despite the wide range of shopping opportunities available to the more mobile population group, proximity is clearly still a potent influence on its shopping behaviour. It is also notable that this is so even for the superstore shoppers. They, along with the residents of the inner-city wards, display a distinct tendency to choose the store nearest to their homes. Distance decay and a neighbourhood effect continue to be significant features of contemporary grocery shopping behaviour (Fig. 7.2).

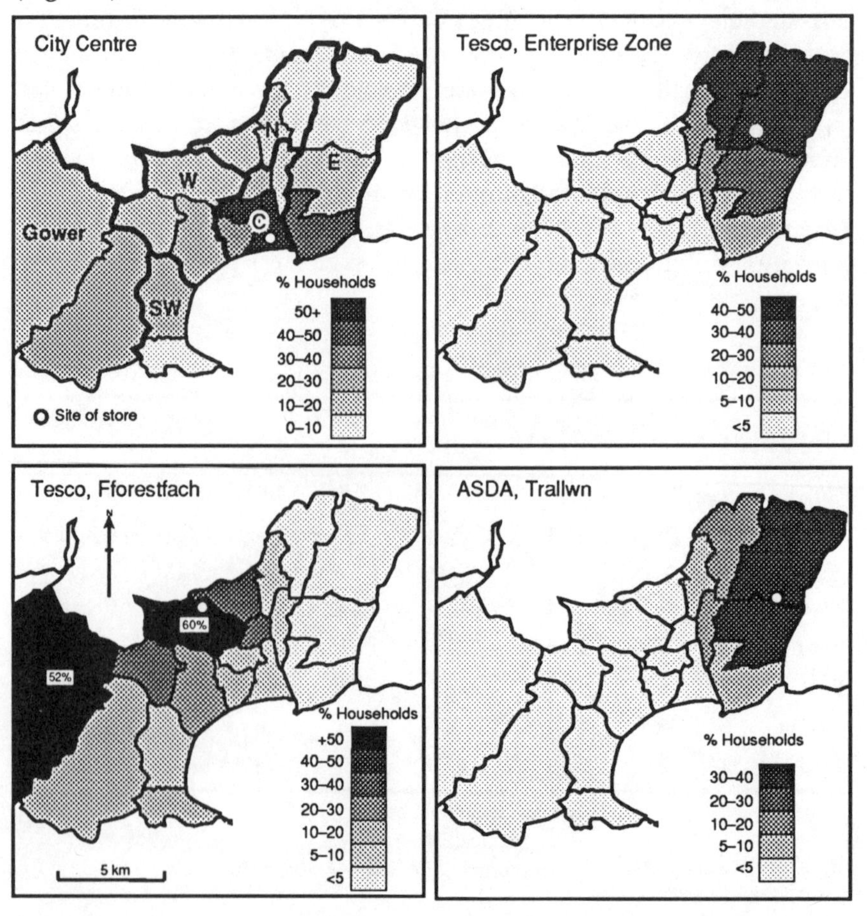

Figure 7.2 Origins of customers at selected main food-shopping destinations in Swansea (by wards).

For those respondents visiting superstores as their main food shopping destinations a supplementary question was asked to ascertain their previous destinations (Table 7.2). This provides a useful indication of the nature of the transfer of trade away from the traditional shopping centres as a result of the development of the superstores. The largest proportion (34 per cent) formerly used the city centre, closely followed by those diverted from the range of district centres (29 per cent), while the smaller local facilities have also been affected to a significant but lesser degree (14 per cent). However, movements away from the Tesco superstore at Fforestfach (11 per cent) and from the Asda superstore at Trallwn (9 per cent) simply reflects the erosion of their early monopoly situations in the western and eastern suburbs before the opening of the other superstores.

The development of the superstores evidently resulted in a significant weakening of the grocery shopping function throughout the traditional shopping centres of the city. It is only in the wards near the city centre and in the sector of the city lacking a superstore (southwest) that the traditional facilities retain an equivalent significance. Clearly, grocery shopping behaviour patterns have been transformed radically by the introduction of superstore retailing.

Table 7.2 Location of previous main food shopping in Swansea (of those using superstores in 1986) (%).

Area*	n	City	Districts	Sainsbury	Tesco	EZ s'store	Leos	Asda	Local
Gower	108	34.3	16.7	0.0	24.1	0.9	0.0	7.4	16.7
South-west	84	27.4	23.8	1.2	28.6	3.6	2.4	11.9	1.2
West	228	42.5	15.4	0.0	10.5	1.3	0.0	6.6	23.7
North	295	20.7	51.5	0.0	6.1	3.4	3.1	6.4	8.8
East	213	29.1	36.6	0.0	2.8	4.7	0.0	16.9	9.9
Centre	196	50.0	12.2	1.5	11.7	0.5	2.6	3.6	17.9
TOTAL	1224	33.6	29.1	0.4	10.8	2.5	1.4	8.5	13.8

Source: Household survey, 1986.
* See Figure 7.2.
Key: EZ = Enterprise Zone; Ffach. = Fforestfach

However, while the changing pattern of grocery shopping behaviour has been demonstrated, it is far more difficult to demonstrate its impact on the functional characteristics and commercial viability of the traditional shopping centres. Ideally, this would require a costly detailed monitoring analysis of retail change throughout the shopping centres of the area, and the findings of such a study are lacking. Nevertheless, a number of exploratory comments can be offered.

The substantial diversion of grocery shopping trips from the city centre of Swansea appears to have had little adverse competitive impact on the functioning of the city centre as a whole. This is not too surprising since even in the late 1970s grocery shopping formed, along with all other regional-level shopping centres, only a minor component of the overall retail turnover of the city centres. Thus, despite the steady decline of many small specialist food stores, the city centre continues to accommodate three well-appointed supermarkets, along with large food halls in two variety stores and an extensive produce market. A similar situation has been noted for Gateshead (Howard 1989) and for Portsmouth (Hallsworth 1990b). In both cases recent reinvestment in supermarket retailing is indicative of a continuing buoyancy in their convenience-shopping rôle despite substantial competition from out-of-centre alternatives.

Of potentially greater significance is the 29 per cent and 14 per cent transfer of trade from, respectively, the district and neighbourhood centres. Food shopping has traditionally been the "anchor" function of such centres, and to a significant degree the other shops located in such centres have depended upon the "spin-off" shopping generated from the grocery stores to enhance their commercial viability. The trade transference associated with the superstores has therefore probably significantly weakened these centres. There is, for example, limited evidence to suggest that the Morriston district centre, in the northern sector of the city, has lost a significant proportion of its former car-borne trade to nearby superstores (Rees 1986). Again, however, there is insufficient evidence to suggest a drastic negative impact throughout the lower levels of the hierarchy. All the district and neighbourhood centres that existed in the late 1970s continue to trade, and retail refurbishment is not an uncommon feature throughout the system.

The scale of behavioural shifts away from the district and neighbourhood centres might be expected to be reflected in their demonstrable commercial decline. The absence of any major effects may well reflect the absence of a sufficiently detailed impact investigation to elucidate the subtleties of retail change. However, a number of additional considerations are likely to mitigate the most adverse local impact. The diffuse and selective appeal of superstore shopping over extensive trade areas has been shown to reduce any localized effects. At the same time, the traditional facilities continue to be used by the less mobile, lower status and elderly sections of the community, and by the remainder for supplementary "top-up" shopping. The polarization of shopping behaviour between the trips exhibited by the more affluent car-owning sections of the community and the "underprivileged" remainder is widely considered to be an important socially regressive impact of superstore development. At a general level there is little doubt that this behavioural distinction exists, although it is not possible at the level of generalization adopted in this chapter to decipher whether the disadvantaged consumer is a significant casualty of contemporary retail change. The issue continues to attract considerable critical scrutiny (Bromley & Thomas 1993) and is developed in greater detail in Chapter 9. In addition, three of the six stores under review - Tesco at Fforestfach, Leos at Penplas and Asda at Trallwn - were developed in part to function as district shopping centres in new residential areas identified as underprovided with this level of services. Developments in such situations are liable to have less effect on existing centres than those in more overtly competitive locations. Finally, retail contraction in older conventional shopping centres is a slow process since established small family concerns will often continue with reduced profits until the proprietor approaches retirement.

However, if viewed entirely from the perspective of the traditional shopping hierarchy there is a danger that the impact of superstore trading will be regarded as largely negative. This is clearly not the case and this form of shopping opportunity is regarded widely as a positive improvement to the urban retail scene. The deflection of bulk grocery shopping trips from city centres is also likely to have delayed the growing problem of traffic congestion and allowed the centres to cater more efficiently for the specialist shopping function which is their chief attraction. In the broader planning context, the superstore has been used widely as a device to enhance the status of conventional middle-order shopping centres. The degree of success of such a strategy, however, depends upon the detailed site relationship between the traditional centre and the new facility. Ideally, a judicious location will promote spin-off trade from the superstore to existing shops, while increasing the access of the less mobile consumers to superstore shopping. This relationship, however, requires more detailed investigation. A recent study in Camden, for example, suggests that the addition of a superstore, although well patronized by pedestrian and public transport-oriented shoppers, has generated an uncomfortable increase in traffic congestion from the car-borne trade (Warren & Taylor 1991).

The proliferation of superstore development in the Swansea area has clearly had a marked influence on patterns of grocery shopping behaviour. For the majority of the population the weekly bulk-grocery shopping trip to the superstore has emerged as the principal feature of food shopping, and the six superstores are a positive addition to the range of shopping opportunities. For a city of this size this level of superstore decentralization has not resulted in dire consequences for any level of the shopping hierarchy. With the continuing growth of car ownership, however, the strength of the traditional centres is likely to be eroded even further, but in the absence of additional large-scale decentralization a reasonably complementary relationship between the traditional and the modern prevails for grocery retailing. On a cautionary note, however, it is possible that if the process of decentralization were to accelerate in the future, the likely increased redistribution of grocery shopping to the new facilities would have a more radically adverse impact, particularly in the middle orders of the traditional shopping hierarchy. This view concurs with that of the Royal Town Planning Institute that: "Smaller convenience stores in local, suburban catchments will reduce in number and will survive only with new capital investment and long opening hours" (RTPI 1988). The point is worthy of reiteration since the question of "superstore saturation" is still far from resolved. The Department of the Environment, for example, recently quoted a commercial view that double the number of superstores trading in 1990 would be viable in the future (DOE 1992a: 22).

The impact of retail warehouses and retail warehouse parks

The initial development of retail warehouses in the late 1970s was considered unlikely to have an adverse effect on existing trading patterns (Brown 1989). In particular, DIY retail warehouses and garden centres were viewed as new retail functions unsuited to trading within the fabric of traditional shopping centres because of their large space and car parking requirements. This view of competition as "benign" fostered the concept of the complementary relationship between the retail warehouse phenomenon and the pre-existing retail system. In the context of the increasingly "free-market" attitudes of central government to retail planning in the 1980s, this view was maintained despite the incremental addition of increasingly specialized retail functions to this form of trading. This was accompanied by the emergence of the retail warehouse park in a variety of forms and scales. The complementary functional view of retail warehouse trading has continued to be reiterated until comparatively recently (Davies & Howard 1988) and is only now being seriously scrutinized (DOE 1992a).

Considering the scale and range of functions associated with retail warehouse trading in the 1990s, and the fact that some retail warehouse parks may well be functioning as new forms of shopping centre, evidence on their trading impact is slight. There is some largely anecdotal evidence of closures in existing centres resulting from the development of retail warehouses. The closure of four furniture stores in Wembley town centre associated with the development of retail warehousing in northwest London, and the relocation of a retail warehouse from Gateshead town centre to the Retail World warehouse park are cases in point (DOE 1992a). Similarly, the development of the Swansea Enterprise Zone retail park was associated with the relocation of four stores formerly peripheral to the city centre and the relocation or contraction of five others from other parts of the city. In total, relocations and contractions totalled nearly 40 per cent of the retail floorspace of the retail park (Thomas & Bromley 1987).

The functional characteristics of retail warehouses and retail warehouse parks are also well known. They serve predominantly car-borne customers so that their trade areas are more akin to traditional regional centres than to conventional centres of equivalent scale. Evidence from the Aireside retail warehouse park in Leeds, for six DIY and furniture stores in Tyne and Wear (Bernard Thorpe and Partners 1985), and the Swansea Enterprise Zone retail park (Thomas & Bromley 1987) reported consistent findings. More recently, surveys by Brown (1989) in Belfast reported similar results. Typically, the great majority (about 85 per cent) of shoppers are drawn from a dispersed trade area of up to 10 miles radius, while the remainder travel from even farther afield. Thus, like superstores, the impact of retail warehouses on conventional shopping centres is liable to be diffuse.

Specific evidence of trade transfer is sparser but is nevertheless consistent with the trade area studies. Evidence from the early 1980s for Tyne and Wear (DOE 1992a) indicated that 27 per cent of shoppers interviewed at six DIY outlets or garden centres had been diverted from Newcastle and Sunderland, while a further 25 per cent named another town centre or local centre in the area.

Further evidence from Swansea

More recent evidence is available for a wider range of retail warehousing functions from the Swansea area. Because of a combination of Enterprise Zone policy and successful planning appeals dating from the late 1970s, a number of retail parks and free-standing retail warehouses were added to the retail system of the area by the mid-1980s (Thomas 1989). Four unplanned retail warehouse parks emerged: in the Swansea Enterprise Zone; at Fforestfach; on the redeveloped dockland site adjacent to but physically separated from the city centre; and at Carngoch, just outside the city boundary (Fig. 7.3). The Enterprise Zone retail park (37,160 m^2, 400,000 sq ft), equivalent to a fifth of the retail floorspace of the city centre, was of near-regional scale, while the remaining three are as large as any of the traditional district centres in the area. A number of free-standing units were also developed at accessible locations throughout the urban area. The large scale of retail warehouse developments in Swansea provides an opportunity to examine the effects of this form of retailing on the traditional district centres in the area.

The Swansea household survey noted in the previous section provided extensive evidence of the impact of retail warehousing on shopping behaviour. Information relating to the last purchase of DIY goods, carpets, furniture, electrical and gas appliances, and clothing and footwear was available, and three distinct patterns emerged (Table 7.3). The data for DIY products closely mirrored the situation for groceries. The new retail warehouses and parks dominated the pattern of purchases (60 per cent), although the city centre was still the single largest destination (28 per cent) and the district centres retained a significant supplementary status (8 per cent). Like the patterns for grocery shopping behaviour, however, the aggregate situation obscures a notable proximity (neighbourhood) effect. The city centre had retained a strong position for the residents of the nearby central wards (49 per cent), despite the absence of a large modern DIY outlet. Conversely, the residents of the eastern and northern wards were far more strongly oriented towards the Enterprise Zone and other nearby facilities (59 per cent and 57 per cent, respectively). By the mid-1980s, therefore, retail warehouses already dominated patterns of DIY purchases in the city, although not for those living in the immediate vicinity of the city centre.

The patterns for carpets and furniture shopping were different. The city centre continued to be the dominant focus despite the development of many out-of-centre alternatives. The city centre attracted 44 per cent of purchases of carpets and 48 per cent of furniture, although the com-

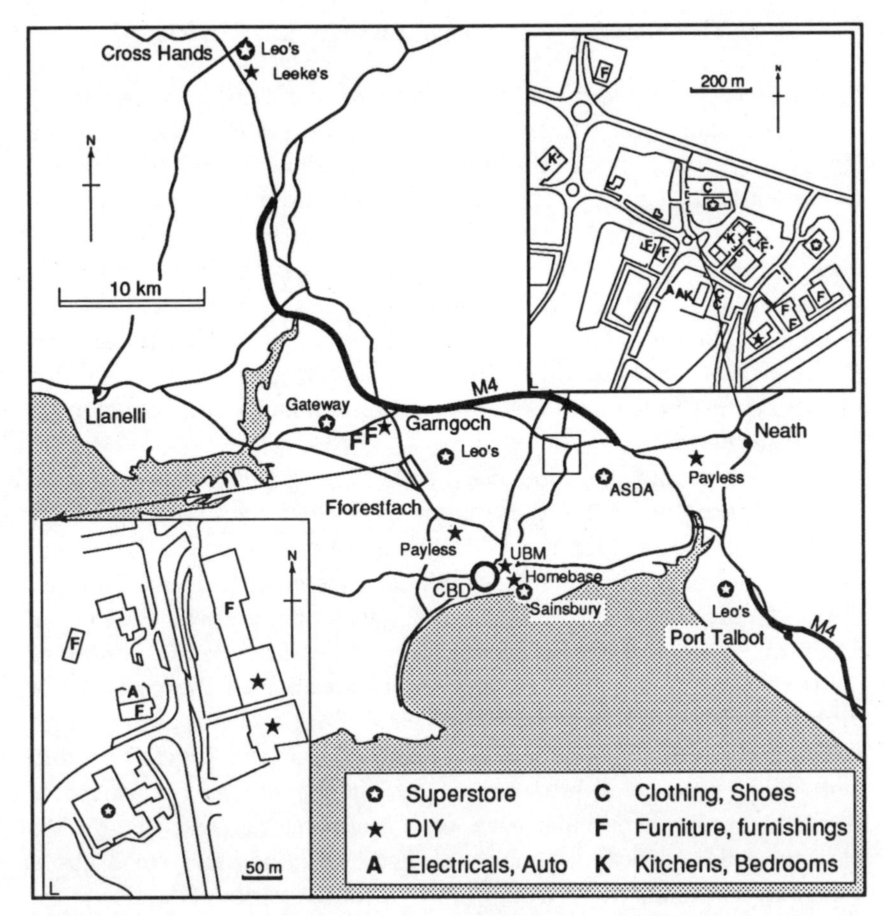

Figure 7.3 Retail warehouses and retail parks developed in the Swansea area, 1977–86.

Table 7.3 Location of last purchase of non-food goods, Swansea shoppers (%).

Goods	n	City	Dist	N'hood	EZ	RWs	Hbase	Superst
DIY	1899	28.1	8.0	1.6	25.2	23.9	5.7	3.9
							other	m ail
Carpets	1868	43.6	6.0	-	15.8	19.9	11.8	2.8
Furniture	1782	47.8	7.4	-	14.3	18.9	8.5	3.1
Electrical and gas appliances	1952	66.4	12.0	-	7.0	8.9	2.2	3.6
						superst		
Clothing or footwear	2104	78.5	9.4	-	5.6	2.5	1.5	2.6

Source: Household survey, 1986.

Key: City = city centre; Dist = district centres; N'hood = neighbourhood centres; EZ = Enterprise Zone; RWs = retail warehouses; Hbase = Homebase; Superst = superstores.

bined effect of the retail warehouses was not far behind (36 per cent and 33 per cent, respectively). The city centre had clearly retained an important carpet and furniture shopping function. Information relating to the early 1970s, however, although not strictly comparable since "main" furniture shopping trips were examined, suggests a substantial decline in the competitive status of the city centre. Residents interviewed in a variety of Swansea suburbs recorded furniture shopping allegiances to the city centre more in the order of 75–90 per cent (Thomas 1974). It is also interesting to note that the minority rôle of the district centres in the mid-1980s was not dissimilar to their situation for the earlier period.

For electrical and gas appliances, and clothing and footwear, a third pattern was apparent. For these more specialized goods the city centre continued to be the dominant location (66 per cent and 79 per cent, respectively), while the retail warehouses were of distinctly minority status (Table 7.3). This reflects the substantially lower degree of decentralization of these functions. Nevertheless, the detailed differential between the two product groups is indicative of the potential of further decentralization, since by 1986 the trend to out-of-centre retailing of electrical goods was more marked than that for clothing and footwear. Three retail warehouses specialized in the full range of electrical appliances, while each of the superstores and DIY outlets incorporated significant elements of electrical goods and small appliances. By contrast, there were only two footwear outlets and one for clothing, although the fact that they attracted 13 per cent of the residents of the eastern wards is at least indicative of the potential effect of the future decentralization of these functions. Comparison of the current data with evidence of the "main" clothing shopping destination for the late 1970s (Penny 1984), however, indicates only a marginal erosion of the strength of the city centre, while the "supplementary" position of the districts remained much the same throughout the period.

Additional findings from the Swansea data are, however, also relevant to an assessment of the emerging impact of retail warehouse trading on the traditional centres. The polarization of shopping trips between the increasing reliance of the more affluent, car-owning consumers on superstores, relative to the orientation of the remainder to the older conventional facilities, can also be detected for retail warehouse trading. For the purchase of DIY materials, for example, 69 per cent of car-owning households used the out-of-centre facilities compared with only 32 per cent of households without cars. The same distinction was noted for furniture: 60 per cent of households without cars had made their last purchase in the city centre, compared with only 42 per cent of car-owning households. Conversely, 39 per cent of car-owning households and only 18 per cent of those without cars had frequented the new facilities. The implication is clear: the traditional centres are becoming steadily more reliant on the less affluent segment of the consumer market, which must, of necessity, undermine their long-term economic health.

The precarious situation of the city centre is also suggested by evidence on reasons for visiting, and the associated attitudes to alternative opportunities. Out-of-centre retailing is particularly attractive to car-borne shoppers for various convenience-based reasons, among which ease of parking usually ranks high. For example 52 per cent of shoppers visiting the Enterprise Zone retail park considered ease of parking to be very important, with a further 39 per cent considering it important (Thomas & Bromley 1987). This overwhelming attraction can be balanced against distinctly ambivalent attitudes expressed by city-centre shoppers. The combined results (1,425 respondents) of two surveys undertaken in the city centre in 1988 and 1989 is illustrative of this effect. Nearly half (49 per cent) of car-borne shoppers disagreed with the view that a parking space could usually be found quite quickly, a figure matched by the proportion disagreeing with the view that parking charges were reasonable. A similar proportion (48 per cent) expressed an active dislike of using multi-storey car parks, while 38 per cent considered that the car parks were too far from the shops and 29 per cent that there were insufficient conveniently located parking spaces. These views do not indicate an imminent collapse of travel by car to the city centre, but neither in the context of the increasing importance of car-borne shopping do they instil confidence.

A number of notable features emerge from the Swansea experience. DIY, like superstore retailing, is already predominantly the preserve of the new out-of-centre facilities. This function is, however, still retained to a significant degree in both the city centre and the district centres by smaller outlets supported by an important neighbourhood effect. This effect appears also to sustain, at an apparently stable level, the continued supplementary status of the district centres for all the more specialized goods reviewed.

Such stability is, however, not the hallmark of the city centre. There appears to be a strong inverse relationship between the degree of decentralization of shopping to retail warehouse sites and the degree of dominance of the city centre for these functions (Table 7.3). This suggests that further decentralization is likely to erode the competitive situation of the city centre even further, while any further reduction in the attractions of the city centre for car-borne trade is liable to exacerbate the situation.

In fact, the status of the city centre is likely to have deteriorated further by 1993. The competitive base of out-of-centre furniture retailing has been broadened by the addition of a number of up-market retail warehouses to the earlier concentration on middle-market traders. Similarly, a number of additional electrical-goods retailers have been established in decentralized sites. Nevertheless, the evidence from the clothing and footwear patterns indicates that the continued strength of the city centre lies in the wide variety of specialist shopping opportunities still almost exclusively offered there.

As in the case of the superstores, the evidence suggests that the retail warehouses have particular appeal for car-borne shoppers drawn from a wide trade area. Thus their impact is diffuse and is likely to affect the town centres and district centres within their spheres of influence. The early notion of their essentially complementary relationship to traditional centres and their benign impact, however, appears in need of re-evaluation. The scale of retail decentralization and the variety of new retail forms associated with retail warehousing suggests the existence of a more potent force for change than was initially anticipated. Retail warehouses are clearly capable of diverting a significant proportion of trade from existing centres for an increasingly specialized range of products. Again, a catastrophic impact on the traditional hierarchy is yet to be demonstrated, although a steady erosion of the strength of the conventional centres can be postulated. However, the trend towards the decentralization of specialist functions continues, as exemplified by the John Lewis Partnership store at High Wycombe and the incorporation of Marks & Spencer and many other multiple retailers in subregional and regional shopping centres. Should such developments be undertaken in the context of the numerous existing retail warehouse parks, creating "hybrid" regional centres (Schiller 1986), then these would be likely to become more directly competitive with existing town centres. The Swansea Enterprise Zone retail park has already gone a considerable way in this direction.

The impact of regional shopping centres

Early proposals for the development of regional out-of-town shopping centres dating from the late 1950s and early 1960s met with vigorous opposition from central and local government. Extrapolating from the US experience, the scale of the proposals for Brent Cross in London and Haydock Park in Lancashire generated fears that city centres would be damaged, urban containment and the green belts would be undermined, and that local traffic congestion in the vicinity of the newly developing system of inter-regional motorways would intensify. In fact, the first regional shopping centre at Brent Cross was not approved until 1968 and did not begin trading until 1976. Even then, it was viewed as exceptional and to fit neatly into a gap in the traditional retail hierarchy in northwest London rather than as a truly new phenomenon.

The early hesitation about such development relaxed with changes in government attitudes to retail planning in the early 1980s. Retail change was given additional impetus by Enterprise Zone policy, which enabled new retail developments to occur in association with economic regeneration in depressed urban areas. However, the subsequent plethora of proposals has continued to be met with official caution. Added to the earlier misgivings has been the fear that in certain cases investment in regional centres might deflect finance from inner-city regeneration. Thus only four have been built, with a number of others still at the planning enquiry stage.

The contemporary situation and the existing regional centres can therefore be regarded as experimental and the evaluation of the centres' impact is likely to have an important effect on future policy. In this context, the small number of subregional centres can be viewed as an interim compromise, with an impact more likely to approximate that of similar sized retail parks than the larger type of development.

Most recent research relating to regional shopping centres has been undertaken by the Oxford Institute of Retail Management and focuses primarily on a large-scale impact analysis of the first of the new phase of regional centres: the MetroCentre in Gateshead which was completed in 1986. The studies adopt a conventional "before and after" survey methodology involving consumers, retailers, land-use inventories and pedestrian flows focusing largely on the MetroCentre and Newcastle. These are supplemented by a telephone survey of the shopping behaviour of 2,000 households in the region. All these sources suffer from the difficulty of disentangling the impact of the centre from other "environmental" changes in the system, but the use of a "collective" perspective on the range of evidence is thought to have minimized this problem.

The MetroCentre, located south of the Tyne, 5.5 km (3.5 miles) southwest of Newcastle upon Tyne, is undoubtedly of regional status. It includes the three large anchor stores of Marks & Spencer, the House of Fraser and Carrefour (now Gateway). In addition, most of the national multiples are represented, together with a number of local independents. The retail floorspace totals 126,480 m^2 (1.36 million sq ft), and there is an additional 9,950 m^2 (107,000 sq ft) of leisure floorspace (Howard 1989). As with the other forms of car-oriented shopping facilities, customers are drawn from an extensive trade area, with 80 per cent spread evenly throughout the 25-minute drive-time zone. This includes most of the built-up area of Tyneside and Wearside, along with much of County Durham. Similarly, mobile, affluent, middle-aged family shoppers are strongly represented, contrasting with an equivalent under-representation of the lower status residents of the inner suburbs and local authority housing estates.

The allegiance of the shoppers interviewed in the MetroCentre for a range of products is indicative of its competitive status (Howard 1989). A high proportion regarded it as their main special shopping centre (43 per cent) and main clothing shopping centre (36 per cent). Similar numbers of MetroCentre shoppers said that Newcastle was their main destination for these functions (41 per cent and 42 per cent, respectively). This suggests that the MetroCentre has made a significant competitive impact on the city centre. However, the high percentages of MetroCentre shoppers continuing to use Newcastle as their main venue for high-order functions suggests that this impact is secondary rather than of the same status as the regional attraction of Newcastle.

Corroborative evidence is available from a survey undertaken in Newcastle that indicated that 75 per cent of shoppers did most of their clothing shopping in the city centre. Also, despite the fact that shoppers in central Newcastle were disproportionately drawn from the less-mobile youngest and oldest age-groups, the most affluent were still more strongly represented than in the MetroCentre (Howard 1989). Newcastle has clearly retained its primary retail status despite a significant competitive challenge from MetroCentre. In fact, it might be suggested that the range of environmental and land-use indicators of the strength of town centres suggested recently by the DOE (1992a: 90) should be supplemented by the aggregate social characteristics of town centre-shoppers as an important behavioral monitor. For example, if town-centre shoppers continue to broadly represent the socioeconomic profile of the wider city region, even when the proportion of pedestrian and public transport-oriented shoppers is increasing, a commercially buoyant situation is indicated.

Nevertheless, the impact of the MetroCentre can be detected in Newcastle. In 1988 and 1989, for the first time since 1976, pedestrian counts recorded declines of 19 per cent and 14 per cent, respectively, following the opening of the MetroCentre. This was reflected in the re-emergence of a "compaction process" initially associated with the refocusing of the CBD on the Eldon Square development after 1976. The contraction of retailing in the peripheral shopping streets, particularly in the southern sector of the city centre, was the most obvious result.

But a seriously adverse effect was only discernible in Gateshead. Pedestrian flows declined by 34 per cent following the opening of MetroCentre, while the retailer survey recorded equivalent reductions in turnover. The decline of Gateshead to a local-district status was suggested, although at this level it was still buoyant and continued to attract investment from major food retailers.

Apart from the focusing effect on Gateshead, substantial negative consequences for the remainder of the traditional hierarchy were not evident. This is consistent with the spatially diffuse and socially selected characteristics of the trade area of the MetroCentre. However of at least equal importance is the fact that the centre transacts only an estimated 8–10 per cent of the retail expenditure within the 30-minute drive time (Howard 1989). This level of turnover is unlikely to have a widespread adverse effect on the viability of alternative centres, particularly if its influence is spread over a large area. Comparably detailed investigations of the more recently completed regional centres are not yet available. However, extrapolating from the MetroCentre evidence a number of qualitative assessments have been made of the likely effects of the Meadowhall centre (111,600 m^2, 1.2 million sq ft) located 5.5 km (3.5 miles) east of Sheffield and 4 km (2.5 miles) west of Rotherham. An extensive trade area is anticipated because of its particular locational advantages. It occupies a highly visible site adjacent to the M1 motorway, with 12,000 car parking spaces and excellent bus and rail access. Compared with the MetroCentre it has 33 per cent more people living with the 30-minute drive zone, and twice the number living in the area up to 45 minutes away by car (Davies & Howard 1989). It is also more accessible to the higher-income areas and has four large anchor stores. As discussed in Chapter 6, the impact of Meadowhall on Sheffield already appears to be greater than that of the MetroCentre on Newcastle.

The situation in Rotherham is likely to be comparable to that of the experiences of Gateshead (Williams 1991). It has for some years been losing ground to the stronger attractions of Sheffield and is considered

35 per cent under-shopped for a town of its size. Proximity to Meadowhall is likely to compound its problems and "compaction" to a convenience-oriented district centre for the less-mobile population is the most likely outcome.

Evidence from Dudley generally supports existing findings (Perkins 1992). The Merry Hill regional centre is 4 km (2.5 miles) southwest of Dudley in an Enterprise Zone location. The initial planned retail warehouse park (36,177 m^2, 389,000 sq ft) has been extended with the addition of a regional shopping centre (111,600 m^2, 1.2 million sq ft) that may compete with Wolverhampton to the north and Birmingham to the east. In the context of the dense network of centres in the West Midlands, Dudley (55,800 m^2, 600,000 sq ft) has functioned as a thriving small town centre for a localized trade area. Its current competitive situation is therefore closely comparable to those of Gateshead and Rotherham, and a similar process of adjustment appears to be under way.

An analysis of the Goad shopping centre plans for Dudley in the five years following the opening of the retail warehouse park in 1986 suggests a decline in the retailing status of Dudley (Perkins 1992). Between 1986 and 1991 the proportion, if not the quantity and quality, of convenience goods stores was maintained and the service functions showed a slight increase (Table 7.4). However, comparison goods declined from 51 per cent to 44 per cent and this was directly reflected in the increase in vacancies. Several important anchor stores relocated to Merry Hill. This was initiated by Halfords in 1986, but the loss of Marks & Spencer, Sainsburys, Littlewoods, British Home Stores and Next to the regional centre in 1989, along with the closure of a local department store (Cooks) has detracted significantly from the former vitality of Dudley. There are vacancies in the primary shopping areas of the High Street, Churchill precinct and Trident centre, as well as the secondary shopping streets, and there is a widespread sense of commercial blight, with many visible signs of neglect. While it is possible that the changes may reflect longer-term adjustments in the wider retail system, store rationalization policy and the effects of recession, it seems more likely from the nature and timing of recent changes that Merry Hill is the central explanatory factor. Dudley, like Gateshead and Rotherham, appears to be declining to a more localized district status for a less mobile clientele.

Table 7.4 Shop types in Dudley town centre 1986–91 (%).

Type	1986	1991
Convenience	14	14
Comparison	51	44
Service	25	27
Vacant	7	12
Miscellaneous	2	3

Source: Perkins 1992.

Reaction in the town centre

So far, we have focused on the negative effects of regional out-of-town shopping centres on CBDs and other smaller town centres in their vicinity. However, these can be mitigated by the redevelopment and enhancement of traditional centres, which can be seen as important positive responses to retail change.

Since the mid-1960s most major cities have designed schemes to maintain their centres as the commercial, social and cultural cores of city regions, while many small town centres and district centres have been radically refurbished, frequently incorporating a new superstore. In the period 1965–89, 604 such schemes of greater than 4,650 m^2 (50,000 sq ft), totalling 8.9 million m^2 (95.4 million sq ft) were developed (DOE 1992a).

Newcastle is a notable example. Howard (1989) notes that the imminent completion of the MetroCentre led directly to the early refurbishment of the Eldon Square centre. This was followed by the extension of trading hours in the city centre, the reduction of car parking charges and the introduction of proposals for new car parking facilities. The 93,000 m^2 (1 million sq ft) Galleries redevelopment scheme for the Bull Ring in Birmingham is a similar reaction to existing and planned centres in the West Midlands (Jones 1989). Even the "compaction" effect is not necessarily negative. If the opportunity is used to change the function of areas peripheral to the retail core and improve their environmental quality, the net result can be the enhancement of the attraction of town centres (see Chapter 8).

Complementary moves to improve access to town centres using land-use and transportation planning devices are also being introduced widely in the conurbations of the UK. The development of inner ring roads and associated multi-storey car parks was an early response. More recently, pedestrianization schemes have become commonplace, and the active promotion of improved public transport has involved "traffic calming", park-and-ride schemes, and the introduction of light railway systems such as Metrolink in Manchester (DOE 1992a). At the same time, a number of major retailers have been promoting the need for "town-centre management" to co-ordinate the planning and promotion of traditional centres. These schemes are still experimental and there are outstanding issues concerning their financing (Morphet 1991). It is nevertheless clear that the traditional town centre is unlikely to become a passive victim of the process of retail decentralization.

New retail facilities can also be used imaginatively to counteract the build-up of traffic congestion in all the major town centres. This is not a new idea. As early as 1968, the Cardiff Development and Transportation Study (Buchanan & Partners 1968) suggested that redevelopment costs in the city centre needed to alleviate traffic congestion could be reduced by the development of two small regional shopping centres in out-of-town locations. Similarly, in 1989 Cambridgeshire County Council included the provision of an out-of-town shopping centre in its structure plan, with the object of reducing pressure in the historic city centre of Cambridge. One of four possible sites is likely to be chosen in 1993. At a scale of one-third of the city centre, each is seen as capable of deflecting pressure from Cambridge, while not being so large that it will seriously undermine the city's primary status (see also Chapter 6). A similar scheme for the Exe Valley at Digby Hospital in the context of Exeter was granted planning permission in 1992.

Caution and an element of ambiguity, however, continue to characterize the debate concerning the future of regional shopping centre proposals in the early 1990s. This reflects the influence of the DOE Planning Policy Guidance Note 6, which directs that planners should not consider the competitive effects of large new centres on the existing system, except where they would seriously affect the vitality and viability of a nearby town centre (DOE 1988). The draft revised guidance of late 1992 has a similar message (DOE 1992b). The commercial orientation of central government, tempered by caution, is epitomized by the conclusion of the Cribb's Causeway (Bristol) public enquiry. The expansion of the existing retail park to a regional shopping centre was allowed on the grounds that there was no convincing evidence to suggest that the development would cause significant harm to the Broadmead shopping area of central Bristol (Updata, February 1992).

The issue of investment confidence in the town centre, therefore, remains an imponderable in the equation (Davies 1991). Levels of re-investment will depend upon the degree to which central government is likely to develop a clear framework for further retail decentralization. A lack of firm direction is unlikely to generate sufficient commercial confidence to stimulate enough re-investment to retain the competitive status of existing centres.

Consistent with the other forms of out-of-centre retailing, the experience of regional shopping centres suggests the central significance of a diffuse hinterland and car-oriented shoppers for their trading pattern. They present a competitive challenge to the traditional city centre that is likely to result in a compaction process on the edge of the CBD. However, the degree of adverse impact will tend to reflect a combination of the existing strength of the centre and its capacity to react to the threat. For smaller centres in the vicinity of the regional centre, the prospect is bleaker, and a decline in status associated with a re-orientation to a more localized market is likely. Positive effects can, however, also be anticipated. The competitive threat of a regional shopping centre is capable of stimulating the revitalization of traditional centres, while the inclusion of regional shopping centres in a planned complementary relationship with existing centres can alleviate the worst problems of inner-city congestion.

International experience

The decline of the downtown city areas in the USA associated with the prodigious growth of regional out-of-town shopping malls underlies the reticence of the approach to this form of retail decentralization in the UK (see Chapter 5). However, the circumstances and scale of decentralization in the USA are not sufficiently similar for the analogy to be directly relevant. The US situation was characterized by lower suburban residential densities and higher levels of car ownership, while the ethnic dimension of inner-city change contributed to the decline in the attraction of the shopping facilities in the downtown for affluent suburban consumers. Lord (1988) notes that by 1982 there were already 14 metropolitan areas in which a total of 25 major suburban retail concentrations exceeded the retail sales levels of the CBD, the trend being particularly marked in Atlanta (3) and Indianapolis (6). In the Baltimore metropolitan area alone (1980 population 2.2 million) there were 10 regional shopping centres larger than 46,500 m^2 (500,000 sq ft), two of which exceeded 92,900 m^2 (1 million sq ft). By 1986 there were 268 centres larger than 1 million sq ft (92,900 m^2), two of which exceeded 185,800 m^2 (2 million sq ft) (Jackson & Johnson 1991). Thus the retail system of the US city has moved from the traditional monocentric form focusing on the original CBD to a polycentric structure in which the original city centre often occupies a commercially secondary and environmentally problematic status.

It is nevertheless interesting to note that even in these adverse circumstances downtown revitalization is possible. In situations where the city centre retains a degree of environmental attraction, advantages of centrality and historical heritage can be used to encourage re-investment. A review by Carey (1988) indicates that this requires a "development function" composed of a joint venture of public assistance and private investment, combined with a strong "management function" to promote positively the city centre. The examples of Boston, Atlanta, Baltimore, St Louis, San Diego and Portland are used to illustrate the process.

The more restrictive planning environment of Canada is considered to provide a better rôle model for the UK than the USA, despite a tendency to allow free-market forces to prevail (Hallsworth 1990b). Federal and provincial constraints on regional shopping centres, along with activities designed to retain the attractions of the downtowns by promoting environmental improvement and public transport systems, have been sufficient to retain varying degrees of complementarity between the city centres and the regional malls rather than competitive conflict. Calgary and Edmonton, for example, have light rapid transit systems to the downtown available free of charge at off-peak periods (Hallsworth 1990b), while the integration of public transport modes focusing on the city centre combined with the development of the Eaton Centre (185,800 m^2, 2 million sq ft) in central Toronto is also noteworthy. For the development of regional centres the essence of the situation is, however, neatly expressed by Gayler (1989b: 280) who argues that provided "they are not too many or they are not too big, they will relieve the town centre of various pressures relating to accessibility and quality of shopping environment".

Defining the precise number, sizes and locations consistent with a complementarity between old and new, however, remains the central problem.

On these issues some interesting evidence is available for Edmonton, Alberta. The metropolitan area (population *c.* 800,000) contains 10 centres of subregional status and a further 6 regional centres of 46,500–92,900 m^2 (500,000–1 million sq ft) (Jackson & Johnson 1991). In addition, located five miles west of the downtown is the West Edmonton Mall, accredited as the world's largest planned shopping centre, and termed a mega-regional centre. It offers just over twice the gross leasable area of the downtown, comprising a comprehensive shopping and leisure centre of 375,000 m^2 (4.04 million sq ft), including five major department stores and the 46,500 m^2 (500,000 sq ft) Fantasyland indoor amusement park (Johnson 1991). It has 23 per cent of the retail floorspace of the metropolitan area and is estimated to transact 42 per cent of the retail expenditure (Fairbairn 1991). The major impact, however, appears to have affected the two closest subregional centres, both of which have been re-created to target the discount segment of the market to reduce competition with the centre.

The scale of the downtown has been clearly usurped, but a drastic diminution of status is not apparent. Partly as a reaction to the development of the West Edmonton Mall, 83,700 m^2 (900,000 sq ft) of retail floorspace has been added to the downtown since 1980 (cf. Newcastle), while the city centre had retained its primacy for the highest order retail functions (Johnson 1991). Evidently, a degree of accommodation between the old and the new is possible even in a situation of substantial suburban decentralization.

The European experience serves to highlight the issue of retail planning. Initially, throughout most of western Europe commercial pressure for retail decentralization in general, and for regional shopping centres in particular, was not resisted. In West Germany there were already 57 regional centres by 1977. In Paris alone, 15 new regional shopping centres were designated in conjunction with a strategy of planned suburban growth. Throughout France this number had increased to 20 by 1989, and there were a further 57 subregional centres (DOE 1992a).

However, since the mid-1970s the early recognition of the competitive impact of retail decentralization, particularly on the city centres, has led to the widespread introduction of stricter controls. This was initiated with the French Loi Royer in 1973 and is also reflected in the gradual strengthening of West German legislation since 1977 (Zentes & Schwarz-Zanetti 1988). Control has been the particular hallmark of the German experience where concerted efforts have been directed at the revitalization of the city centre with an emphasis on environmental enhancement and public transport strategies (see also Chapter 8). At the same time, where decentralization is seen as advantageous to the reduction of inner-city congestion, the new retail forces have been accommodated in the emerging strategies. However, experience in the Netherlands is portrayed less optimistically. Borchert's (1988) review suggests that a profusion of legislative controls has not been very effective and this has precipitated major problems of inner-city decline, particularly in the largest centres of Amsterdam, Rotterdam, The Hague, Utrecht and Eindhoven.

In general, the European evidence promotes the value of firm direction, designed to reduce uncertainties and to accommodate commercial innovation. In the UK, this view has been promoted by Davies (1986) since 1986 and it recurs throughout a recent review of retail charge in the UK (DOE 1992a).

Conclusions

In the UK much is known of the functional characteristics of the many new out-of-centre retail forms and their impact on changing patterns of shopping behaviour. The superstores have transformed and now dominate patterns of grocery shopping. The early idea that retail warehousing was focused on a new range of functions and, therefore, had a benign effect on the traditional hierarchy has had to be revised as free-standing stores have been replaced by retail warehouse parks of steadily increasing scale and functional specialization. A combination of solitary retail warehouses and retail warehouse parks has been shown to be capable of eroding the strength of conventional district centres and even central business districts. Evidence of the effects of retail warehouse parks also suggests the potential impact of their possible gradual transformation into hybrid regional centres in a "quiet revolution"; or of the latent impact of additional subregional centres. Evidence from the few regional shopping centres also indicates that they constitute a significant competitive challenge to city centres and can seriously reduce the status of nearby smaller town centres.

Negative impacts

The current scale of retail decentralization has clearly caused a general weakening of many traditional centres. The deleterious results of these changes on the traditional system of shopping centres, on access to shopping facilities and on traffic circulation has generated widespread comment. The negative impacts can be summarized into three points:

- The decline in the competitive position of many existing centres of all levels and their associated problems of environmental deterioration has caused considerable concern to traders and public agencies alike. Also, in a limited number of cases, the development of a new facility near to an existing, weak centre of similar functional status has resulted in the drastic decline of the older centre.
- The inability of the less mobile sections of the community to make full use of the new out-of-centre shopping facilities has also given rise to misgivings, particularly by local planning authorities.
- Despite the much-vaunted commercial advantages of the accessibility of the new facilities, they can generate traffic congestion if the issue of traffic circulation is not addressed adequately.

Positive impacts

The retail revolution appears to have been accommodated by a new complementarity in which the old and the new continue to function in tandem, albeit unevenly. In fact, it is possible to identify three principal advantages of the new retail facilities:

- They offer a wider range of shopping opportunities than previously existed which has a strong appeal for the car-borne population right across the spectrum of shopping behaviour.
- The competitive threat posed by the new facilities has acted as a stimulus for the refurbishment and revitalization of some traditional centres.
- While the new opportunities generate more and longer car-borne shopping trips, they may have deflected traffic from the town centres and contained, or even reduced congestion.

Prospects for the future

The evidence reviewed in this chapter suggests that the impact of further retail decentralization will be dependent upon a number of factors, including scale, location, response, and planning control.

Of these, the issue of scale is central. The evidence from the superstore sector, for example, suggests that further large-scale decentralization will have more marked effects on the remaining alternative facilities. Similarly, the unconstrained development of regional centres on the North American model is capable of producing comparable polycentric and environmentally problematic results. In the absence of reliable predictive models, the logic of existing studies of retail impact suggests that relating the scale of a new proposal to that of the existing system is the most useful point of departure, if viewed in the light of the functional characteristics of the variety of new retail forms. The question of location is of similar importance. The new facilities clearly continue to have most effect in their immediate neighbourhoods. At the same time, response within the existing system is of major consequence. The conventional centres continue to be attractive to substantial sections of the community and the enhancement of such attractions can both strengthen and expand this appeal.

In looking to the future, however, it is important to stress the advantages to be gained from the minimization of investment uncertainties combined with firm planning controls over the process of retail change. The ideal scenario would maximize the advantages accruing from the development of new facilities, while minimizing the negative economic, social and environmental consequences of the decline of the conventional centres. In the contemporary political climate, central and local government direction that reduces uncertainties for investment is considered a necessity (DOE 1992a). For maximum effect such a policy will need to allow for the accommodation of commercial innovation in retailing rather than focusing on a restrictive policy of containment.

CHAPTER EIGHT

Planning and shopper security

Taner Oc & Sylvia Trench

This chapter focuses on the planning aspects of security problems faced by shoppers and other users of city centres. It argues that functionalist planning causes security problems and that other planning strategies are needed to make cities safer. It will review some examples of successful policies and practice from recent experience in the UK and the USA.

The principles of functionalism set out in "the Charter of Athens" (Conrads 1970) required rigorous zoning that clearly segregated work, home and leisure. The roots of functionalism can be found in the garden cities movement, and modern British planning after the Second World War was based on functionalism which has been more extensively practised in the UK than in other European countries. Functional separation of activities and land uses led to the construction of transport networks to link them. Then in the 1960s, influenced by studies like the Buchanan Report, many cities put a tight noose of ring roads around their centres. Thus the centres were not only deprived of their complexity of activities but were further isolated as single-function areas.

As the recent Green Paper of the Commission of European Communities (CEC 1990: 26–30) states: "Functional separation may some time be useful when applied, for example, to industry. In other areas, however . . . the practice of strict zoning ignores the patrimony and geographical reality of the city. 'Functional exactness' destroys the flexibility of the city and its buildings; these, conceived as architectural objects, are unable to adapt to changing conditions and therefore prevent the city from functioning as a dynamic whole." Problems created by functionalism in planning have been widely discussed (Jacobs 1961, Sennett 1991) and some writers elaborate on the safety aspect of environments created in post war-planning (Valentine 1990, Roberts 1990, Wilson 1991).

After the Second World War British cities started cleansing their city centres of uses deemed incompatible with the image of industrial cities, which were supposed to look prosperous, confident and aggressive. Birmingham was a good example of the desired image. The blueprint for the post-war city centre was Coventry, which was completely redesigned and rebuilt after extensive bomb damage. On the other side of the Atlantic rising CBD land values and transport developments had resulted in a similar erosion of the complexity and vitality of city centres, shaping them into islands of activity alive only from 9am to 5pm. In the 1960s and early 1970s in the USA it was said that even muggers went in threes in the city centres after dark.

The same problem of deserted city centres is beginning to emerge as a major problem in the UK as more out-of-town shopping centres are built. The basic requirements for successful retailing are maximum visibility, accessibility and security. Lack of security in city centres means loss of trade, which may set in motion a downward spiral. Experience in the USA shows that the fear of crime impedes the ability of retailers to serve both existing and potential customers. As a city centre declines, office workers are less inclined to shop during lunch hours (Hassington 1985). As the mobile, higher-income shoppers desert the declining city centre, the area becomes dominated by lower-income and ethnic-minority shoppers. Higher-income shoppers prefer to use retail areas used by people like themselves (Citizens' Crime Commission 1985) and they feel safer in streets with similar types of people.

This study, carried out in New York in 1985, argued that "an economically vibrant downtown [city centre] concentrates in a geographically compact area a wide spectrum of business, social and leisure activities. When a large number of people can walk quickly and safely from one activity to another, a downtown assumes its two unique characteristics: the capacity for visitors to engage in multi-purpose visits, known as the multiplier effect, and a high level of interpersonal communication" (Citizens' Crime Commission 1985: 12). The study (ibid: 13) showed that fear of crime damages a city centre's economy by inhibiting the behaviour of its users in the following ways:

" ○ it depresses the multiplier effect by reducing the level of pedestrian activity and the distances people are willing to walk on the streets;
- ○ it encourages insulated activity in which self-contained complexes and indoor walkways are preferred to outside sidewalks;
- ○ it decreases the level of face-to-face communication between users;
- ○ it promotes the desertion of the area after 5 pm;
- ○ it increases automobile use and demand for close-by parking."

The fear of crime robs a city centre of those elements that make it a unique retail location as people avoid either coming to it altogether, especially when there are a growing number of out-of-town shopping malls, or limit their activities significantly. Either way, retail and leisure outlets become the first losers. The New York study showed that in the USA people are afraid of being mugged, raped or assaulted, and of having their cars broken into or stolen. As many as 63 per cent of the respondents said they rarely used the city centre because of fear. Similar attitudes to visiting the city centre are beginning to emerge in the UK. A 1989 survey carried out in Nottingham by Research Services of Great Britain interviewed 987 people aged over 16 about their fear of crime, perception of risk and avoidance behaviour (Nottingham City Council 1990). This revealed that a substantial proportion of respondents avoided the city centre after dark, and of those who did use it in the evenings, 45 per cent said they felt unsafe. Some even avoided the centre in the daytime through fear of crime – 3 per cent on weekdays and 7 per cent at weekends. About half those questioned said they were worried about mugging and being attacked by strangers. The most common reasons given for feeling unsafe were "people hanging about" (57 per cent) and a general fear that "something may happen" (38 per cent). The respondents mentioned 27 specific places that they would avoid, including Market Square, Nottingham's main central square.

Because city centres are perceived, especially by women, to be dangerous they are either deserted or given over to gangs of revellers and elderly drunkards after dark. This not only denies large numbers of men and even greater numbers of women the use of their city centres at night, but also has a significant economic and employment cost. This could be of the order of £24 million per annum and over 600 jobs in a city like Nottingham (Nottingham City Council 1990).

The British government has seen crime and fear of crime as major contributors to the problems of declining urban areas, and a Safer Cities Programme formed part of its 1988 "Action for Cities Initiative" designed to promote social and economic regeneration in problem inner city areas (Home Office 1988). The connection between safety and prosperity is clearly underlined by the statement that creating "safer cities where economic enterprise and community life can flourish" is one of the three main objectives of the programme. The Home Office is now supporting projects in 20 towns, cities and London boroughs chosen because they have high crime rates together with other economic problems.

Avoiding areas identified as potentially dangerous is a rational human reaction and many people given choices do not use city centres after dark. However, there are significant numbers of people, many of them women, who have to be in the centres after dark or before 9am as a condition of their employment, notably the large army of female cleaners. Indeed, women as shop assistants, secretaries and shoppers are probably the dominant presence in the city centres. It is necessary to find ways of bringing them safely into the city centre, and it is also necessary to implement measures to make city centres safer for them by day and night even though it is recognized that the potential of physical changes to improve the situation is intrinsically limited.

Strategies sometimes known as situational crime prevention can only ameliorate what is termed opportunistic crime: those crimes committed by the rational criminal who weighs up the opportunity for gain against the cost of being observed, physically prevented or caught (Newman 1972). Even in these cases the effects are limited; there are many instances of robberies and assaults in relatively well lit and public places. Design can only create the preconditions for a safer environment: it is a poor substitute for changing the conduct of the offending individuals. But in so far as some planning policies can affect either actual crime or perceptions of safety, planners have a responsibility to explore what contribution they can make. It is to such planning measures to create safer city centres that this chapter will turn next.

Design and development solutions

There are three main strategies for dealing with the kind of fears noted above. The first, tackling the root causes of crime, is outside the remit of this chapter. The strategies that do concern planners directly may be divided into two distinct approaches: first, the provision of single-purpose, segregated and protected malls with guards. This is the "fortress approach", which has significantly contributed to the deterioration of the city centres in the USA and is now coming to Europe. The second measure is to improve the city centre to make it safer. This will have a multiplier effect on all the activities of the city centre and not just benefit the retailing sector (Davies 1982, PPS 1984, ULI 1983). The rest of this chapter will look at measures for creating a safer city centre for retail and other activities. Many of the best examples are drawn from the USA, which has experienced some of the worst city-centre security problems but also demonstrates some of the most successful remedies.

Cyril Paumier's *Designing the successful downtown* (Paumier 1988, 1982) notes seven basic strategies essential for "reshaping the space-use composition and economic vitality" of city centres:

- promote diversity of use;
- emphasize compactness;
- foster intensity;
- provide for accessibility;
- build a positive identity;
- ensure balance; and
- create functional linkages.

These seven principles are instrumental in building safe environments in city centres.

Housing and mixed-use development

Some American city centres came back to life during the 1980s with the growth of gentrified areas adjoining city centres. The proliferation of restaurants and other features of yuppie life styles allowed them to function after dark. Market forces are not likely to produce similar results in the UK, so there is a need to consider what central and local government can do to make city centres safer. Arguably the most important policy change to be considered must be how to bring housing back into city centres and make them 24-hour zones again. Experience in the USA has demonstrated the value of this policy and it is also supported by experience in most of continental Europe, where housing was never pushed out of the centre to the extent that it was in the UK.

It is well established that people feel safer in the presence or visual range of others (Valentine 1990, Jacobs 1961). City-centre residential areas are an effective way of creating well used streets and providing natural surveillance. Even during the 1960s and early 1970s when many city centres in the USA were deserted at night, one part of Philadelphia city centre, Rittenhouse Square, continued to be a safe and elegant oasis surrounded by a diverse neighbourhood of mixed housing, clubs, art galleries, restaurants, offices, a music school, a few small workshops, a cinema and some expensive shops. By contrast Washington Square, which was surrounded by tall office blocks, was totally deserted after 5pm in spite of expensive redevelopment.

If planning policies are to be used to make city centres safer, the whole basis of zoning policies needs to be reconsidered. Ways of creating city-centre residential areas must be explored; for example, incentives might be given to developers to invest in city-centre housing like the tax incentives given to businesses to create jobs in Enterprise Zones. The recent "Living above the shop project" could be very useful in bringing back residents to city centres if more funds were made available (Boseley 1992).

Increased surveillance can be achieved by two measures. First, land uses can be mixed and, secondly, streets can be designed in such a way that there is density of activity and clear visibility over long distances. But getting more people onto the streets requires that the streets themselves have a greater variety of activities available. A report prepared by COMEDIA (1991) draws attention to those of the city centre activities that were zoned out by planning or pushed out by land prices that are the very kind of activities most attractive to families, middle-aged men and women and the elderly. If such people could be brought back into the centres it might be possible to reclaim them from the domination of drunk and disorderly youths.

Women shoppers feel very unsafe and although crime statistics do not show them to be particularly vulnerable, statistics are misleading. There is much under-reporting; much of the behaviour that deters women is more a matter of unpleasant harassment than acts that are technically criminal, and the very fact that women stay away reduces the number of possible attacks on them. In any case it is women's perception that influences their behaviour and surveys show that large numbers are afraid to be out in the dark and some feel unsafe during the day as well (Atkins 1989). Bringing back certain activities like swimming pools and sports facilities and keeping them and public libraries open for longer hours would ensure that larger numbers of people use city centres over a longer part of the day. Also, staggered hours would mean that even shops could be kept open longer, as is the practice in a number of other countries. In most British cities even the public library, which is the only place used by both sexes and different age groups, is often closed early in the evening. Pubs and fast-food chains dominate the city centres in the evening and they only cater for a small group.

In the 1980s a number of US cities experienced a dramatic change in the fortunes of their central areas, with significant increases in the number of city-centre residents. The pattern started after 1973 with gentrification and has gained momentum with the development boom. The most spectacular example is in Chicago. North Michigan Avenue, the shopping street for the residents of the area known as the Gold

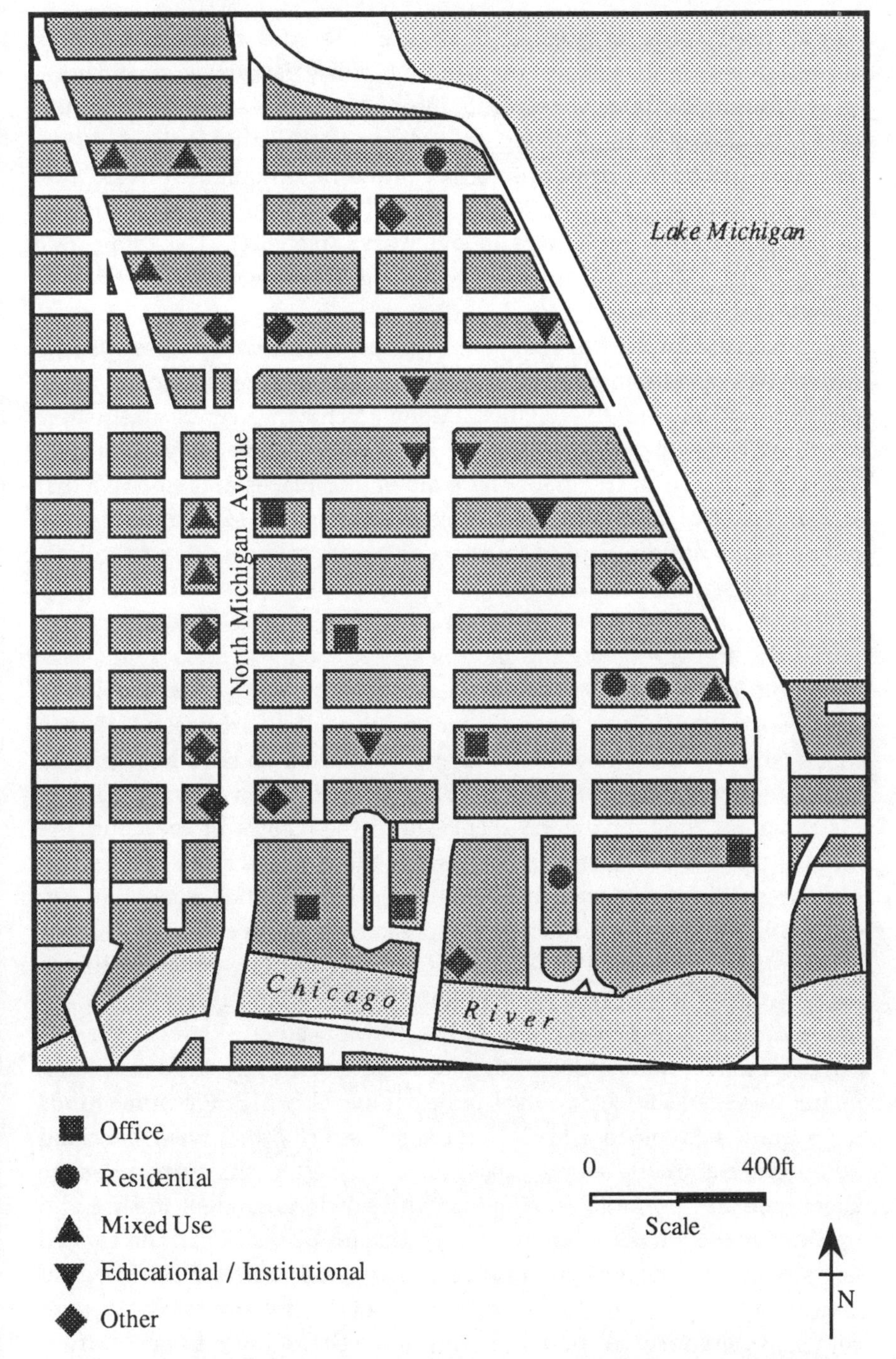

Figure 8.1 Chicago: downtown development in the vicinity of Michigan Avenue, 1989–92.

Coast, used to be a muggers' paradise after the shops and offices closed. But recent changes in planning policy favouring mixed uses have brought 20,000 new residents into the area known as the Magnificent Mile and its immediate vicinity (Mazur 1991, Fieller & Peters 1991). The area has been turned into a 24-hour zone occupied by residents, tourists in hotels, shoppers, diners, as well as office workers, and is one of the safest areas anywhere in the world (Fig. 8.1). The crime rate along North Michigan Avenue is among the lowest in the city of Chicago.

The success of North Michigan Avenue is not solely based on its residential developments, it is also an excellent example of what planners call an activity corridor. Along Michigan Avenue itself there are three large up-market shopping malls where shops stay open until 9pm, and in two of the malls there are eating places and cinemas that stay open until midnight. The streets running off Michigan Avenue offer literally hundreds of restaurants and cafés that also stay open late.

Another significant factor in making the area safe was the cleaning-up of Rush Street; the sleazy girlie bars are gone, the go-go dancers and porno theatres are gone, and in their places are more restaurants, jazz clubs and discos, all of which are used by both sexes. The drunkards have moved away from Rush Street and the whole area around North Michigan Avenue is now much safer. It must be noted that while the undesirable uses are no longer in Rush Street they have not altogether disappeared; they have been displaced to the fringes of the Loop. As long as there is a demand, activities like these will relocate. However, planning policies can make sure that at least substantial sections of city centres are free of such functions and safer for shoppers.

One of the cities in the UK that is implementing a policy of mixed uses by trying to increase the amount of residential accommodation in the centre is Wolverhampton. It has introduced a "Living over the Shop" scheme and has also designated a part of the city as an entertainments quarter. The city council is providing Urban Programme funds for a grant scheme to attract restaurants, street cafés, wine bars and other entertainment-related uses to this quarter. It is intended to encourage and promote on-street activities including street theatre and music to make it a pleasant and lively place for pedestrians. The council has already taken steps to improve the physical fabric of the environment and to promote the reoccupation of empty upper floors over shops. A new resident population in Wolverhampton's town centre is seen as essential to making the area self-policing and overcoming people's fears. A variety of funds, including the "City Challenge" fund will be earmarked for improvement schemes that provide living quarters over shops in the area. Some housing associations are also interested in being involved in the scheme (Scarth & Ashcroft 1992).

The surest way in the long run to create safer urban environments is to create zones of mixed land use with a substantial amount of residential accommodation. However, this change might take a long time, especially in UK cities, partly because of cultural preferences rooted in suburban life styles. In the meantime, other measures like creating activity corridors can be used to improve safety. In fact, safe activity corridors can also act as catalysts in establishing residential areas in city centres.

Dense and compact development

Housing and mixed-use developments provide the functional basis for a high level of activity in the city centre. The second measure designed to ensure activity can be achieved by planning dense and compact developments that will increase the concentration of pedestrian movements. A busy and attractive core area minimizes the fear of crime. It is necessary that there is at least a core area of the city that respectable people tend to frequent. The use of pavements, cafés, shops, and offices by middle- and upper-middle income groups in large numbers creates a positive secure image that acts as a basis for expansion. Milder (1987) notes that the median length of a Manhattan shopping trip is only 1,200 ft and that 75 per cent of all pedestrian trips are of less than 2,000 ft. Therefore, "the diameter of a core area should usually be well under one-half mile" (Schwartz 1984).

There are several examples of downtown revitalization schemes in the USA that are notable success stories. Michigan Avenue North in Chicago was first established as a safe activity corridor so that developers could be attracted to develop residential units that they would be confident of selling. The Magnificent Mile is a perfect example of an activity corridor: an avenue that is well lit, has varied land uses, a long uninterrupted sightline, and area management policies that ensure a relatively high level of activity until after midnight. The only area the police have to keep an eye on is around the West Water Tower where visibility is interrupted.

Another notable example is the 16th Street Mall in Denver, a mile-long spine running from the and Civic Centre to Lower Downtown

(Fig. 8.2). "Flanked by retail stores, office buildings and civic spaces, it represents Denver's counter attack on suburban malls and their free parking" (Collins et al. 1991). It is also a compact core retailing area with a very high level of daytime activity, which has created a safe and pleasant environment. Parking lots are in close proximity as they would be for a suburban mall but, unlike a suburban mall, the street is the hub of the public transport system. The provision of a free tram service adds further to the level of activity and distributes it along the street. One very important feature of the 16th Street Mall is that the retail enclaves developed along it are designed with open or wide-glass frontages. The enclosed spaces are visually an integral part of the sidewalk so that there is a constant feeling of a busy street even if some of the shoppers are on the other side of a glass panel.

A compact and secure urban shopping centre is exemplified by the Martin Luther King Jr Center in the Watts district of Los Angeles. For maximum shopper security the developers have surrounded the site with an eight-foot-high wrought iron fence, installed security cameras and floodlights, and built an observatory, as well as having infra-red detectors at the six entrances (three for cars, two for service, one for pedestrians) (Buckwalter 1987). The centre is designed for maximum pedestrian movement in a compact area. In other words, because of the extreme violence and insecurity in the area, many of the features of an enclosed suburban mall are replicated in this shopping centre. Shoppers can come protected in their vehicles, park in a secure car park and shop in safety. However, this is not an alternative to safe activity corridors in city centres, as it is mono-functional. Shoppers tend to prefer safe multi-functional centres.

One mistake in the early attempts to create compact city centres with dense activities was to build limited access structures like the Trump Tower on Fifth Avenue in Manhattan or the Water Tower Place on North Michigan Avenue in Chicago. These buildings have turned their backs on the streets and created well controlled pleasant interior space that is in competition with the street, not part of it. Such structures keep people from using normal pavement routes, especially if they are connected by off-street pedestrian networks, as is the case in Minneapolis or Melbourne. They diminish a city centre's street life and prevent visitors from engaging in more than one activity. One retailer is quoted as saying that "they rob from the street" (Citizens Crime Commission 1985). By contrast, the two new structures with limited access along North Michigan Avenue have opened their ground floors so that they are part of the street system. They have several entrances

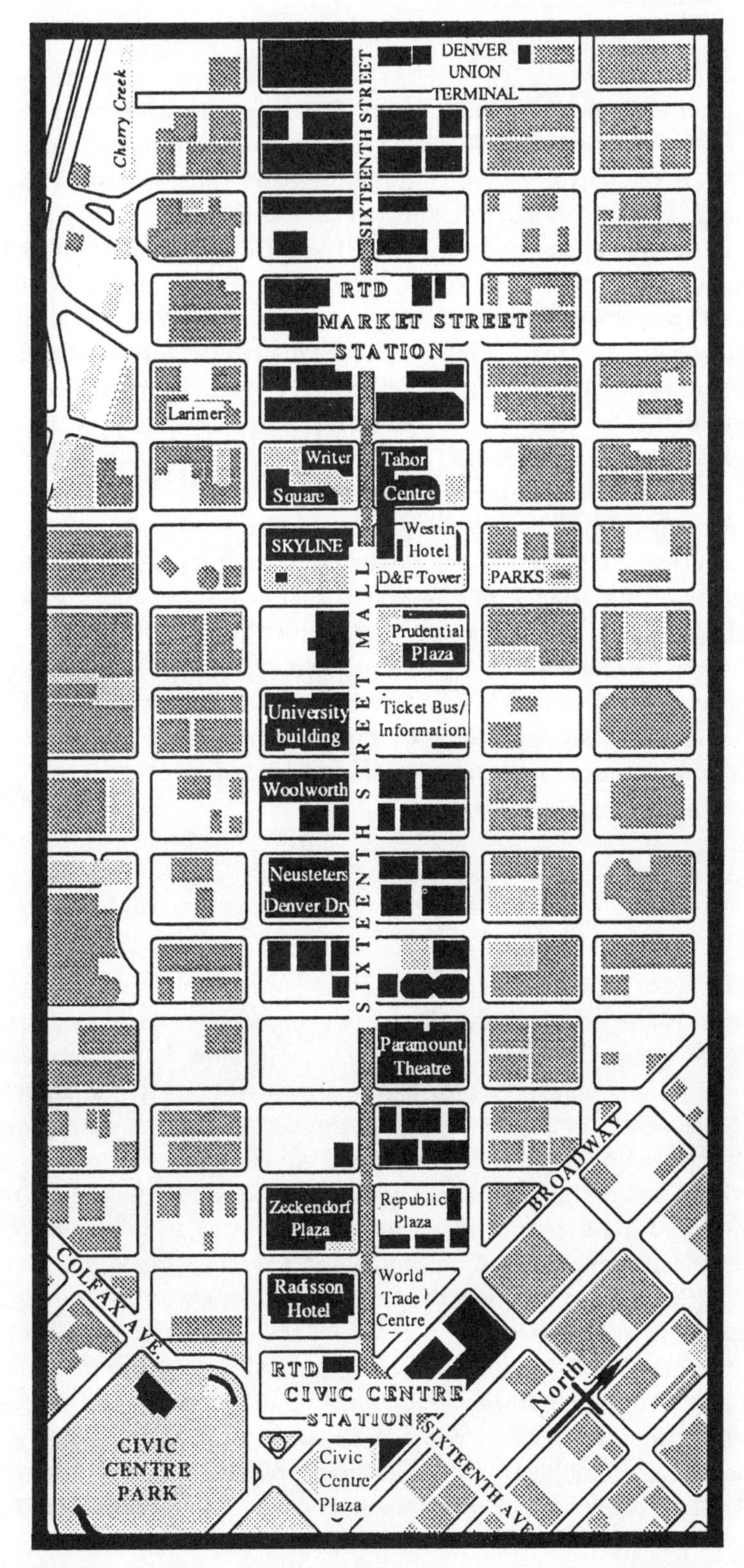

Figure 8.2 Denver: redevelopment for safe shopping in the 16th Street Mall area.

and ground-floor shops that enhance the street with their display windows. Another example of a good limited-access development, in the UK, is the Victoria Centre in Nottingham, which forms a magnet and an enclosed safe street extension during the day.

In designing dense and compact city-centre developments to create intensity of use, the critical mass of activity should be complemented by a balanced distribution of activity generators. A balanced distribution would require that there were evening as well as daytime activities, and weekend as well as weekday activities. Overconcentration of office development, for example, leaves streets empty not only in the evenings but also at weekends (see also Ch. 5). This is why retailing in areas dominated by offices is a weekday only activity.

Some cities may choose to concentrate their safety measures on a few central activity corridors. This policy has certain benefits, since those using the central area would be concentrated in these corridors giving a feeling of security to shoppers and other users. However, such a policy has the implication that other parts of the centre may be even less safe. Wherever they are implemented, activity corridors will only work if co-ordinated with the provision of public transport services, car parks and increased provision for on-street parking.

Positive identity

Bringing back land uses like housing and sports facilities to city centres takes time and requires significant policy changes. Activity corridors could be designed and implemented sooner but not immediately. However, some measures may be implemented quickly and with very little cost. First, improved lighting creates an immediate feeling of safety. Using bright white lights instead of yellow street lamps brings a significant and relatively cheap improvement. Another measure to make streets more inviting and attract greater activity is an improvement of the physical appearance of buildings and keeping streets free of litter and graffiti. Research shows that people are very sensitive to signs of deterioration in the shopping areas in city centres (Citizens Crime Commission 1985, Milder 1987). In the UK, a number of councils including Hammersmith and Fulham, Camden, Leeds, and Leicester, have guaranteed to remove racist or sexist graffiti within 24 or 48 hours (National Association of Local Government Women's Committees 1991). The Urban Programme enabled many cities to clean their old buildings, and they are now making sure that they stay clean.

City-centre events are often used to build a positive image and to attract residents who do not normally use the city centres. However, if these events take place in the early stages of improvement programmes they often have a negative effect, confirming in the minds of many that the centre is dirty and unsafe.

Access and urban design

Parking

Policies towards car parking need to be reviewed in the light of safety considerations. There is a case for allowing on-street parking, especially in the evening. The car is perceived by many as a relatively secure means of travel and many people who have access to cars say that they would be more willing to come to city centres if they could park on the street close to the wine bars, restaurants, theatres, etc that they use. Until recently adherence to the principles of functionalist planning resulted in segregation to avoid pedestrian/vehicular conflict and design to ensure minimum interruptions to vehicular mobility. Inner ring roads and pedestrian subways were also built to satisfy these aims, with cars being left at multi-storey car parks on the fringes.

Multi-storey car parks present many problems. They are normally unstaffed, visibility is poor, and lifts and stairs for people arriving and leaving are out of sight of traffic, both in and outside the car park. Design improvements that increase visibility are possible, as shown by examples from Nuneaton and Leicester (Atkins 1989). Birmingham City Council Department of Planning and Architecture conducted a questionnaire survey of over 500 women on the design and safety of car parks (Birmingham City Council 1990) and a car park design guide was produced in 1992 that incorporates many safety features.

The presence of staff has been shown to reduce crime and increase feelings of security. Some women favour the designation of women-only car parks or the reservation of ground-floor spaces for women near the attendant. This system works well in a number of German cities but when Birmingham City Council tried to introduce a small central area women-only car park it was advised that this could contravene equal opportunities legislation and the proposal has been shelved. Women's groups consulted about women-only car parks without attendants have been divided about whether they should be supported (Trench et al. 1992).

Most people consider well lit, attended, open parking lots to be more secure than multi-storey car parks. This was recognized in the choice of open parking lots for the 16th Street Mall in Denver but not in the developments in Chicago, which provide multi-storey car parks in the centre, and open lots only on the edges which are not very safe. In recent years traffic-calming measures have made it possible for people and vehicles to share the use of central areas with increased safety for pedestrians; the next development needs to allow increased on-street parking at night and to make car parks secure.

Safer public transport

City centres could be made more accessible by putting money and effort into public transport improvements so that people can feel safe and comfortable when travelling on ordinary services. The kind of measures required includes increasing staffing at stations, providing conductors on evening buses, extending hail-and-ride minibus services into housing estates and taking all passengers very close to their homes. The provision of benefits additional to the improvement of safety, combined with the extension of assistance to groups other than those worried about attacks would also help by increasing the general level of usage.

However all these strategies require much more public expenditure in a sector that is already underfunded and under pressure, and even if the money were to be found the public system would not be as secure as a door-to-door service. The special safe women's transport services that have been introduced in a number of cities to provide a segregated door-to-door service usually run only in the evenings and so are not available for shopping trips. Taxis might be the ideal way for women without a car to have a door-to-door service and their use is increasing even among lower-income groups. However, recent cases of attacks on women passengers by taxi drivers and those posing as them may inhibit this growth. It is unfortunate that the development of cab firms using women drivers to carry women passengers is, like the Birmingham car park, running into trouble with equal opportunities legislation. It should be a matter of priority for all concerned to find a way of exempting a development that is potentially so beneficial to both users and city traders.

Not all the measures are expensive. Moving bus stops nearer to shops or petrol stations where there is some evening activity, is a simple measure. Changing evening bus services to hail-and-ride on less densely used routes has been successful in Stockport. This has not required any special financing apart from some expenditure on publicity, while the new service has generated extra travel revenue (National Association of Local Government Women's Committees 1991). On some London Underground stations passengers are being given the opportunity to wait in well lit entrances where there are ticket collectors, only going to deserted platforms when trains are announced. Tyne and Wear Metro run shorter trains in the evening to concentrate passengers, thus protecting them from the dangers of empty compartments. A number of cities in the UK are planning new light rapid transit systems, and their plans should make provision for secure travel with well sited stops to bring shoppers into the city centres.

Subways

Pedestrian subways and barriers to stop people crossing at street level were part of a policy dating from the 1960s that valued motorists' time more than the convenience of elderly and laden pedestrians. In recent years, a re-evaluation of these policy priorities along with safety considerations have resulted in a widespread review. There are ways of improving the design of subways to increase visibility (Atkins 1989) or shops might be located to make them less deserted. However, complete abolition is generally preferred and brings benefits to all those using cities on foot. A Nottingham City Council planning policy review opts for the more radical policy of replacing subways with pedestrian surface crossings (Nottingham City Council 1990). This follows decisions to close pedestrian subways in Glasgow and proposals for future closures in Sheffield, Portsmouth and Poole.

Conclusions

A number of cities around the world are taking steps to make city centres safer and are preparing design guides (Leicester City Council and Leicester Constabulary 1989) and policy documents (Toronto City Council 1988, Nottingham City Council 1990). These documents concur that the following principles should be applied in city centres:

- a mixture of land uses to increase the level of activity and create safer environments;
- natural surveillance of the full length of streets to deter opportunistic crime;
- co-ordination of the safety of people and property (high security fences and blocked-off rear doors and windows increase building security but reduce safety for pedestrians and employees);
- good maintenance not only to eradicate neglect but to prevent buildings and other amenities giving off signals of neglect and insecurity to passers by;
- provision of well used, well lit corridors of activity.

Some of these principles have already been adopted in Chicago, Denver, Toronto, Leicester, Nottingham, Birmingham and other cities. These principles can only have a lasting impact if changes to bring back greater complexity to city centres as outlined earlier are also implemented by cities.

This chapter has argued for high levels of activity but this is a necessary not a sufficient condition of shopper security. It is also necessary that certain undesirable activities are excluded. It is the exclusion of undesirable activities that makes the Magnificent Mile and Rush Street in Chicago such pleasant and safe areas.

It is also necessary to find ways to bring shoppers safely to these areas. In this respect again 16th Street Mall, which has ample parking facilities and good public transport access, provides an example of good practice. By contrast, the Magnificent Mile somewhat fails in terms of parking.

Cities in the UK can benefit from the experience of Chicago and Denver in terms of reclaiming central areas and can also try to bring residents back to city centres through incentive-bonus schemes. They must also start creating activity corridors without undesirable uses so that more people lift their self-imposed curfews and start using city centres, especially at night.

In conclusion, it is important to stress that many shoppers perceive city centres as unsafe and thus avoid using them. This means that many residents in our cities are reducing their choices and narrowing the scope of their activities. As a result the retail sector is losing revenue and declining, particularly in the worst-designed shopping centres. Moreover, there is a danger that this may be the start of a downward spiral. Planning is at least partly responsible for the changes that took place in our city centres in the post-war era through its functionalist policies. Changing to multi-use zones and bringing back the complexity of urban life increases opportunities for natural surveillance. This in turn reduces opportunistic crime and people's fears. Activity corridors reinforce this effect. Lighting and improved environmental conditions bring immediate benefits in terms of perceptions of safety in the central areas.

However planning and design can only create the preconditions for safer environments (Valentine 1990), and as the report by COMEDIA (1991) stresses there is a need for behavioural changes if city centres are to be made safer for the public. Young men in the UK get drunk in pubs and look threatening whereas young men and women on the continent socialize at bars and cafés, very seldom becoming drunk and disorderly. When city-centre facilities are shared by both sexes a safer environment is created. It is partly behaviour patterns that make the centres of most continental cities and Chicago's Magnificent Mile safe, even at night. Another factor is a low-key police presence, which makes a major contribution to shoppers' perceptions.

PART FOUR

SOCIAL ISSUES OF RETAIL CHANGE

CHAPTER NINE
The disadvantaged consumer: problems and policies

Tim Westlake

The retail revolution has brought material advantage to most of the population through greater choice, comfort and cheapness in shopping. These benefits are greatest in superstores, but such outlets have a highly uneven distribution. As a result of strong disparities in income and mobility, there is a marked inequality in consumer opportunities and the existence of "disadvantaged consumers" is now recognized as a major social issue. The characteristics and implications of this issue are developed in this chapter. The disadvantaged include low-income families, women, ethnic minorities, the elderly and the disabled, all of whom share the common characteristic of low mobility. Planning policies adopted to combat the problem include better public transport provision, the protection of local shops and improved design and facility provision in shopping centres. Investment in the new alternative of Electronic Home Shopping (EHS) is a recent approach to planning for disadvantaged consumers.

Since the Second World War, the number of retail outlets has declined dramatically, while the real volume of retail sales has increased. Wrigley (1988) states that there have been two major results of this trend: a significant increase in the size of stores and a dramatic increase in the concentration of capital. The concentration of retail activity into larger decentralized units and the development of large-scale convenience stores offering a wide variety of low-price convenience goods has been a response by the private sector to the demands of a highly mobile society with the ability to buy and store in bulk. As a consequence, certain consumers have been forced to shop in declining local supermarkets and shops that are unable to compete, in terms of price and choice, with the larger decentralized superstores.

Accompanying this movement in retail activity has been a trend to the switching of capital, population and employment out of traditional central shopping locations. This has resulted in a major loss of retail demand from the urban core areas and the creation of large, relatively underprovided concentrations of demand in the suburban, outer suburban and urban fringe areas of Britain (Wrigley 1987). Rees (1987) argues that planners often assume that the growth of large out-of-centre stores is a result of retailers responding to the requirements of the majority of the shoppers. In practice this may not be the case; some of those making trips to car-borne shopper locations are doing so only because facilities have declined in traditional centres.

In recent years a variety of literature has been published on disadvantaged consumers, by authors such as Bowlby (1979, 1984, 1985, 1988), Davies & Champion (1980), Guy (1980, 1984, 1987), DOE (1992a), the RTPI (1988), and Westlake & Dalgleish (1990). Davies & Champion (1980) identified two groups of deprived consumers. First were the "disadvantaged" consumers, a large minority of the population with collectively low income levels and limited purchasing power, who were constrained to use mainly local supermarkets and other small shops. They were estimated to comprise 25 per cent of the population, with large numbers concentrated in the inner city areas. Within this broad group Davies & Champion identified various subgroups of inadequately served consumers. These include the elderly, large young families, unskilled manual workers, the unemployed, the sick and infirm, and those without cars.

Secondly, Davies & Champion identified a small minority of "neglected" consumers, seen as those suffering extreme disadvantage. This group included handicapped people, the elderly with severe mobility problems, families with large numbers of children, and families with bedridden relatives.

More recently, the RTPI (1988) has identified those groups in society most likely to be disadvantaged in terms of shopping as low-income earners, residents in areas with poor public transport, those without cars, people with caring responsibilities (usually women), the elderly, the disabled and those with mobility problems, the young, and ethnic minorities.

Women can be added to these lists as they still tend to carry the burden of convenience shopping in most households and yet have less mobility than men.

This chapter discusses the characteristics of disadvantaged consumers, with a particular focus on women, the disabled and the elderly. It then looks at some of the policies that have been developed to alleviate the problems of disadvantaged consumers. Finally, it examines new technology in retailing, and specifically EHS, which are identified as a potential solution to the access problems of many consumers.

Disadvantaged consumers

Disadvantaged consumers are constrained in their access to a wide variety of stores. Bowlby (1979) studied accessibility, mobility and shopping provision in Oxford. Access, she argues, does not describe the actual usage of shops, nor the cost of travel, rather it describes the ease with which shopping opportunities can be reached. Access is a function of a person's "potential mobility", i.e. the ease with which a person is able to travel if he or she wishes to. "Actual mobility" is used to describe the movement actually undertaken. Bowlby (1979) identifies three important factors that influence a person's "actual mobility": car ownership, income level, and health and personal ability, with each variable interrelating. This shows that consumers must satisfy certain criteria before benefiting from superstore shopping. Rees (1987) indicates that varying accessibility is a "major cause of social inequality", although little analysis about the effects on disadvantaged consumers is available.

In the 1940s, residential areas and shopping centres were planned with primary access for pedestrians. More recently, the car-oriented organization of land use has caused a division in society: there are shops accessible to all and some only accessible to car owners, meaning a "widening of the gulf between the independently mobile and those dependent on public transport" (Rees 1987: 6). Car ownership and use is not diffused homogeneously among the population – 38 per cent of households do not own a car (Pickup 1988) – and is particularly likely to be lacking in disadvantaged groups. As the average weekly shop for a household of four weighs 70 lbs (RTPI 1988), access to a car is important. This problem is often exacerbated by poor public transport links. Bowlby (1985: 219), summarizes the phenomenon felicitously, pointing out that "whilst the fewer and larger superstores offer benefits, it is the ability to reap the benefits which is presently largely confined to the more affluent and physically mobile people".

Although larger city centres endure much new competition, further down the retail hierarchy there are high vacancy rates, store losses, smaller units and pressure to alter uses. Invariably, the net losers are the inner cities, with declining populations and low-income families. Low-income families find bulk-purchasing virtually impossible and often have fewer facilities for transporting their shopping. Price comparison is vital for families with small weekly incomes and affordability is often a more important consideration than the quality of food.

Ethnic minority groups have specific shopping problems. Ethnic minorities tend to have higher than average unemployment levels and large numbers of low-wage earners. These groups have traditionally concentrated in inner cities and tend to have special shopping needs that are usually only catered for in smaller-scale inner city stores. The RTPI (1988) argue that ethnic minorities should be viewed as a group of consumers who require special provision. This provision tends to be provided by specific ethnic and religious groups, which make a unique contribution to the range of shops available in the UK. As ethnic minorities now constitute about 4 per cent of the UK population, it is essential that their needs are in some way catered for (RTPI 1988).

The UK's rural population also has special retailing problems. Rural areas are well documented as suffering deprivation in three distinct areas: household, opportunity and mobility (Phillips & Williams 1984). Rural populations, especially in remote areas, have limited employment potential. Improved mobility acquired through car ownership has increased consumer choice, although that choice is in direct competition with public transport and village stores. The RTPI (1988: 35) states that such stores "play a vital rôle in the social life of the village and the loss of a shop has been seen as one of the signs of a village ceasing to be a community and becoming a dormitory".

Guy (1980) argues that rural shopkeepers have suffered because of increased consumer mobility, declining rural population and increased difficulties in obtaining goods from wholesalers. McLauglin (1986) showed that village shop closures were on the increase and had a disproportionate impact on the disadvantaged. He proved a mutual interdependence between poor households and local shops. With the closure of village stores, disadvantaged consumers, with limited mobility, then incur additional expenditure and the considerable inconvenience involved in travelling to alternative shopping facilities. This has a particular impact on the elderly, and on young families with children, especially if they do not own cars.

Guy (1980) highlights the plight of the disadvantaged consumer by arguing that those consumers who most need the price savings offered by superstores are those who have least access to superstores. Furthermore, by reducing prices to the more mobile and well-off shoppers, superstores are widening the disparities between the rich and poor. Retail corporations ignore those without cars, who, in their opinion, are dwindling in numbers and have comparatively low purchasing power; yet Bowlby's household survey in Oxford (1975–6), indicated that 16–25 per cent of the population under 65 lived beyond walking distance of a supermarket (Bowlby 1979). Thus, the profit-maximizing ethos has nurtured the material advantage of those who have most to spend.

Deprivation, where "members of society do not enjoy benefits which they feel are due to them" (Eversley 1990: 13) is a relative concept; thus consumer deprivation is endured only in relation to the majority of consumers. Eversley (1990) identifies three situations faced by disadvantaged people:

- the greater the concentration of disadvantaged people, the less the odds of breaking out of the poverty trap;
- economically disadvantaged regions, with concentrations of disadvantaged people, have less chance of change for the better;
- the lower the level of incomes, the more the informal or "grey" economies depress the probability of benefits or wage subsidies.

Linked with these situations, stagnant centres, where retailing predominates, may have less money for regeneration and a smaller variety of goods, while being dependent on the patronage of an increasingly deprived clientele. Although these comments relate to particular spatial situations, most discussion of disadvantage refers to particular groups. Here it is proposed to focus on three such groups in order to highlight the diversity of the problem, and the consequent difficulties faced by public and private agencies.

Women

Bowlby (1984, 1985, 1988) has shown that almost 80 per cent of households in the UK are made up of married couples with or without children. Within married households it is generally the wives who carry out the bulk of everyday household shopping and they have the major responsibility for organizing household shopping trips and expenditures (Bowlby 1984, 1988). This has remained the case even though 60 per cent of married women of working age are in paid employment (Bowlby 1988) and 54 per cent of mothers (Bowlby 1984). This tradition of female responsibility for shopping has constrained shopping time. But Bowlby

(1985: 219) argues that women are also anxious to get value for money, "both because they are shopping on tight budgets and because shopping effectively is important to their self esteem."

Rees (1987) in turn states that these factors have led to an increase in bulk-buying, a decrease in the number of shopping trips, and a greater peak at lunchtime, in the early evenings and at weekends. With rising living standards since the mid-1970s, private transport has become increasingly available, so that a greater proportion of the population relies on more occasional bulk-purchase trips for most convenience-shopping needs, facilitated by the massive increase in refrigerator and freezer ownership. These trends have had a profound influence on the locational preferences of retailers because of the rise in independent mobility of consumers.

Women have been disadvantaged in their access to mobility. Hillman et al. (1974) were among the first to show that household car ownership is not synonymous with individual access to a car. The British National Travel Survey of 1985/86 (Department of Transport 1988) indicated that 62 per cent of households owned a car, yet only 41 per cent of women had a driving licence compared with 74 per cent of men. Furthermore, evidence from a number of studies shows that two-thirds of women residing in car-owning households rarely have a car available to use whether they have a licence or not (Pickup 1988). As a consequence of this, women are more reliant than men on public transport for shopping.

Pickup (1984) argues that women's low mobility is a product of their gender rôle and refers to the varying degree of obligation attributed to domestic tasks. Mothers without access to cars may be discouraged from travelling because they may be in charge of young children and may need to face the difficulties of taking prams on public transport. Pickup (1984) states that only one-third of women have personal use of a car, and many are on a strict budget, seeking value for money. Many traditional shopping centres are poorly designed, for example, subways and isolated centres deter apprehensive women, threatened by fear of sexual harassment. Bowlby (1985) argues that convenience shopping by women often results in little social contact, long queuing, limited seating, and problems in amusing children, which together make convenience superstore shopping a tiring and aggravating chore.

The disabled and elderly

Disabled people have been traditionally described as disadvantaged in terms of shopping, yet little research has been undertaken to identify their needs or difficulties. The disabled are not a homogeneous group, although it is possible to identify three characteristics that many of them share:

- a limited ability to travel far without motorized transport;
- a limited ability to use current conventional means of transport; and
- low incomes.

Disabled people tend to have limited mobility, which makes shopping difficult. The problems and preferences of the disabled, in relation to retailing, are frequently ill recognized or analyzed on a partial basis. About 6.5 million people, or 9 per cent of the population, in the UK are to some degree mobility handicapped (Imrie & Wells 1992) . Mobility handicaps may include the long-term problems of those with wheelchairs, as well as temporary difficulties or other physical, sensory and mental impairments. These disabilities may involve considerable strain, stress and pain in moving around, yet people with disabilities are "often prevented from carrying out normal daily tasks, such as shopping, because of inaccessible buildings" (Access for All 1990). The mobility range of the disabled is further limited by obstacles, such as kerbs and steep slopes.

Elderly people are often identified as having similar characteristics to the disabled. This group consists of a fifth of the 20 million households in the UK (RTPI 1988). The elderly, many of whom rely on government pensions, need to be seen as a priority when considering accessibility to supermarkets. Many elderly people dislike shopping in superstores, finding them "tiring, confusing and unfamiliar" (Bowlby 1985) and are loyal to local stores that offer personal contact. Three-quarters of all disabled people are elderly, while poor health further compounds shop mobility.

For both elderly and disabled groups to shop independently they require easy access to shopping facilities, not only in terms of location, but also store layouts. Stores often lack room for wheelchairs, have no ramps to enter and their checkouts are not designed for the disabled. On the other hand, given their low incomes, the elderly would benefit from the cost savings available from large stores.

Solutions to the problems of disadvantaged consumers

The factor linking disadvantaged consumers is their restricted access to retail facilities. The diversity of groups identified limits the solutions, many of which are financially unviable. The remainder of this chapter outlines some solutions to relieve the problems of disadvantaged consumers; EHS, in particular, is identified as a possible, future, large-scale solution to consumer access to shopping facilities.

The previous section indicates that the problem of consumer deprivation is in part the result of commercial aspirations. These have led to decentralization and concentration of retail activity. The main tool for intervention in retailing is planning policies. Guy (1980) argues that planning policies should be considered as a process of intervention by the public sector in the operation of the free market. This model is of particular relevance to retailing, where the market consists of a supply side (retailers and developers) and a demand side (the consumers). Planning policies can attempt to alter the behaviour and decisions of both sides to compensate for any imperfections (Davies 1985). The rôle of the public sector should therefore be to ensure that the benefits of superstores are maintained and social costs minimized. The development of strategies to ensure an equitable distribution of benefits is constrained by the need for superstores to be located in economically viable situations.

Planners are constrained by planning legislation, local government financial resources and a plethora of advice from the DOE, which has argued that "it is not the function of planning to inhibit competition among retailers or among methods of retailing nor to preserve existing commercial interest as such" (DOE 1988).

A more optimal solution would be for retailers to adopt strategies directed at eliminating consumer deprivation. Since the public sector must operate within strict budget constraints, private-sector initiatives allow greater scope for policy approaches. The deprived sector of the population has considerable latent demand that could provide a new market for retailers and improved welfare for deprived consumers. Increased numbers of elderly people over the next 20 years will mean that this latent demand will grow.

Davies & Champion (1980) stress the need to deal with each section of the disadvantaged separately. They argue that specific commercial investments should be promoted and should involve fully researching the areas of need and catering for these needs directly. Whatever the approach to the problem, there seems no doubt that society must intervene in the marketplace to protect the large minority of disadvantaged consumers.

There would be a less profound divide in society if more stores were accessible to those who do not own cars. A revision of policy could help disadvantaged consumers improve their accessibility, for example, by locating stores near isolated peripheral estates via land-purchase agreements with the council. Unfortunately most car-biased superstores have "at best a token bus service linking the nearest town centre" (Roberts & James 1990). A policy change in the fare structure of UK public transport should favour short-distance travellers over the more wealthy commuters who travel longer distances from the suburbs. Those travelling on foot, bicycle or public transport could travel with immense ease if these travel modes were given serious government consideration. Guy & Wrigley (1987) call for improved foot access to local centres by better integration of pedestrian systems. Bowlby (1979) suggests that women's mobility could be improved by an attempt at rôle reversal; retailers could encourage male shoppers and family shopping through advertising. She also argues that locating superstores or outlets close to places of work, such as near the central business district or on industrial estates, might lead to time savings. Conversely, employers or unions could supply minibuses for shopping. Women-only carriages on trains and metros and women's taxis could help to alleviate some apprehension.

A report by the Royal Town Planning Institute (RTPI 1988) stipulates policies that the various actors in the retail industry should undertake. Central government, for example, is encouraged to develop experimental access areas for disabled shoppers. It argues that local authorities can improve access for the disadvantaged by special consideration in policy preparation and in determining planning applications. Retail developers are encouraged to liaise with local authorities to consider access difficulties; suggestions include free bus services and subsidized home shopping. The report also focuses on improving vehicle and station design through negotiations with public and private transport operators. Facility provision can be implemented by amending building regulations in favour of disadvantaged consumers, for example, by replacing steps with ramps for the disabled and altering the design of facilities like telephones, toilets, seating and cash dispensers, so that the less mobile can gain unimpeded access. Facilities for the disabled should be the result of detailed considerations, so that, for example, space can be allocated for manoeuvring wheelchairs on the basis of the reach dimensions of wheelchair-bound people. Further guidelines on disabled access would be expedient, such as the need for clearly marked, reserved parking bays, at a maximum of 40 m from shopping centres, with adequate space between bays for car doors. Pedestrian routes should be free of obstacles like kerbs, with a gradient of not more than 1 in 20 and a vertical level of not more than 500 mm (Access for All 1990).

RTPI (1988) sensibly suggests that local authorities should identify potential and accessible retail sites in their local plans. However, none of their accessibility policies indicates where disadvantaged consumers actually are; perhaps local authorities should use criteria such a car-ownership figures or mobility rates to target their policies. The population census can provide ample information to delineate where consumers are most deprived, thus enabling the focusing of aid.

Many disadvantaged consumers can benefit from new facilities and planners should consider this in the detailing of planning applications and preparation of strategies. Local authorities could implement design standards to ease movement within the store, such as trolleys for the wheelchair-bound, seating, toilets, crèches or covered facilities. Facility provision could target particular disadvantaged groups, such as pedestrian access for women with children.

RTPI (1988) recommends targeting the retail and development industries to improve design and facility provision. It is also argued that initiatives like shop-mobility schemes could help to alleviate the disbenefits of store access for the elderly and disabled. Ultimately policies should encourage the provision of facilities for multi-purpose trips, for example, the location of post offices, chemists, libraries and community facilities in close proximity to each other and proposed or existing shopping developments. Facilities for young children are still often neglected and the provision of both crèche facilities and child-minding would aid mothers.

The government can have considerable influence over retail developers without excessive expense, for example under section 106 of the 1990 Town and Country Planning Act, local authorities can enter into an agreement with any person interested in the land and secure "planning gain". As Moore (1990) asserts, local authorities have the "opportunity for obtaining a 'planning gain' for their community" by negotiating with developers to contribute towards infrastructure or public amenity costs. Likewise planning applications could be given priority where disadvantaged consumers benefit. However, as Thompson (1990: 11) notes, "distributional matters should always be taken into account and given weight in plans, projects and control decisions".

New superstores, for example, should only be tolerated out of town if a small outlet, selling fresh produce, is opened in the inner-city areas. It is not unreasonable to give back to the public a share of the multinationals' enormous profits. RTPI (1988) recommends that local authorities give preference to retail sites in or near population centres in their local plans, as ad hoc provision does not meet the needs of the disadvantaged.

Developments showing an awareness of disadvantaged consumers are not unheard of, as Thompson (1990) notes: disabled access and access by public transport are regarded as legitimate planning issues in some appeals. Impact assessments might be viewed with less scepticism were a separate body to undertake them, rather than a local authority, which is often accused of bringing about a predetermined outcome. New superstores should be assessed in terms of socioeconomic characteristics rather than the likely impact of sales turnover on existing stores.

New alternatives

The Royal Town Planning Institute report (RTPI 1988) argues that the introduction of information technology in retailing, and EHS, will have "only limited impact on retail distribution systems" this century. The development of EHS, which involves the sale of goods and services through interactive viewdata systems, could also dramatically change the shopping habits developed over the past 20 years (Taylor 1984, Westlake 1990, 1992). EHS is a relatively new and developing area of retailing, and one in which technological innovation is constantly changing. However, in broad terms two main types of electronic shopping distribution channels exist: videotext and television programming (Manchester University, 1988). "Videotext" is the generic name given to systems that transfer information, in the form of words, numbers and graphics, to a television screen at the touch of a button. Any information that can be stored in a computer can be transmitted to remote terminals, including adapted household televisions, either through direct broadcasting or over telephone or cable networks.

The second type of EHS is television programming. For a number of years home-shopping companies have advertised items on television with a telephone number for ordering, although such advertising has been limited by Independent Broadcasting Authority regulations. It would seem likely that future developments in both cable and satellite television will set up the base for an expanding home-shopping market, with mainly the sale of comparison goods advertised on television and ordered either through Prestel or by telephone.

Cable television is viewed by home-shopping companies as a major sales medium, given the success in the USA, where there is a home shopping network offering a 24-hour "shopping show". Unlike videotext, cable television displays the merchandise adequately on screen, has the advantage of being interactive, and therefore facilitates quick order-processing. The number of homes receiving broad-band cable has expanded dramatically since its introduction in 1984. The cable authority in the UK reported that over 250,000 homes were connected to cable by the end of 1988 (Retail Business 1988).

Satellite television has the same potential as cable television, and has grown steadily since the merger in late 1990 of the Sky and BSB satellite television services to form British Sky Broadcasting (BSkyB). In early 1991, 1.4 million UK homes were receiving BSkyB channels (*Financial Times* 11 March 1991: 6). There has been continued speculation that these figures will increase steadily for the next five years, and although these have been hit by recent financial problems, the future still looks relatively optimistic (Nisse 1991). The number of people watching satellite television in a total of 11 European countries almost doubled in 1989 from 24 million to 43 million, watching on average seven hours a day (Wittstock 1990). It would seem likely that the UK will follow this trend.

As of 1992, the UK involvement in EHS has been little more than experimental and based on Prestel. Prestel was developed as a means of utilizing spare off-peak capacity in the telephone network, and its original target was the domestic market. When it was launched in 1979, 100,000 users were expected by the end of 1980; in fact there were 10,000 and a majority of those were business users (Bennison 1985). A small number of specialist applications were successful, for example, travel and financial services, where large amounts of quickly redundant information needs to be available to many people. A variety of local projects have been developed in recent years, although some have subsequently ceased after experimental periods. Two types of service have developed: first, community-based schemes, providing services to the elderly and housebound; and secondly, commercial projects.

Community schemes

Community schemes have essentially welfare goals, and tend to be coordinated by local authority social services departments. Such schemes are aimed at giving the relatively housebound access to supermarket goods at supermarket prices. As a result of these welfare goals users have not had to pay installation or usage charges (Davies & Reynolds 1988). Community schemes have tended to be far more successful than commercial projects, probably because the aims are not profit-led, and as such are not necessarily commercially viable.

The first UK project was the Gateshead Shopping and Information Service (SIS), set up in 1980 by the local council, Tesco Plc and Newcastle University. It provides home-shopping and community-information services on Prestel and is aimed at the disadvantaged, particularly the elderly, disabled, single-parent families and the housebound. Terminals were originally located in homes, community centres and other focal points, and food is ordered and delivered at no extra charge. Today terminals are located in the social services department and orders are received by telephone. Gateshead SIS has achieved its objective, to improve the ease with which the relatively housebound can undertake their shopping on the basis of a wide variety of information and with increased choice. This system has 1,000 regular customers, but further expansion is seen as a problem because of the rapid escalation of costs above this level (Davies 1985, Davies & Reynolds 1988). The scheme is still a great success, with a waiting list of potential subscribers (Davies & Reynolds 1988).

Since then, two further community-based systems have developed. First, Bradford Centrepoint was set up in 1986, with around 500 customers receiving free delivery and terminals located in a variety of localities (Retail and Distribution Management 1986). This system was modelled directly on the Gateshead SIS (Davies & Reynolds 1988), but when the Labour lost the 1988 local government elections, the newly elected Conservative council cut its funding to Centrepoint to the extent that it was subsequently closed. There is no evidence to suggest that the closure resulted from the failure of Centrepoint to meet its objectives. Instead, closure merely reflected the different financial priorities of the new council.

A second community-based system was set up in March 1988 by Asda on the Isle of Dogs in the London Docklands. The scheme has 1,000 users, and terminals have been located in sheltered-housing units and community day centres, with goods again being delivered free of charge (London Docklands Development Corporation 1988, Retail and Distribution Management 1988).

Commercial services

A variety of commercial services have also been developed. For example, Club 403 was launched in December 1982 as an experiment to assess the demand for EHS and to produce a commercially viable package of services at an acceptable price (Davies & Reynolds 1988). The service was based in the West Midlands, attracting 1,500 members using Prestel. Customers were forced to purchase their own Prestel equipment and, when using the shopping service, payment including a delivery charge was made on delivery of the goods. Research was undertaken on Club 403 by the Department of Trade and Industry (DTI), which found that of the 1,500 members, over half utilized the armchair shopping services, which were the most popular part of the scheme. Much of the shopping service was based around Carrefour and food retailing, and users were permitted access to Carrefour's national on-line product file (Davies & Reynolds 1988). For the first two years it was subsidized by the DTI and British Telecom (BT), but once funding was ceased the experiment was not considered worth continuing and was terminated in June 1987.

A further commercial example is Telecard Supershop, which was set up in 1985 to provide grocery provisions, with ordering and delivery services, to clients within the inner boroughs of London. The service operated through the Lalani chain of convenience stores, and customers placed orders from an extensive range of grocery and household goods via their Prestel equipment. Goods were more expensive than in supermarkets but cheaper than in traditional corner shops. Delivery was made to a customer's door and payment was made immediately by cheque. Telecard identified target households within five London boroughs, which had high numbers of economically active households and Prestel users (Davies & Reynolds 1988).

Data on Telecard Superstore is hard to obtain, but Davies & Reynolds (1988) state that they failed to meet their planned household penetration, which was 1.4 per cent of households by the end of the first year. This was despite being located in one of the densest concentrations of young ABC1 households in the country. It may be that food shopping alone was an insufficient trigger to those not already using Prestel. Davies & Reynolds (1988) argue that the complex variety of costs and forms associated with the service probably outweighed any eventual benefits. The scheme has subsequently closed down, having failed to attract a large enough customer base.

The use of EHS in the UK seems likely to remain limited, at least in the near future. Since 1979 BT's Prestel system has attracted 100,000 subscribers, in contrast to the French equivalent, which has attracted 5 million (Tilley 1989). France has achieved this massive user population in part by initially donating terminals to potential users. BT decided against a similar policy and has stated that what "the French have done is excellent, but we have a different philosophy. There are merits in both cases. BT is continuing a philosophy that providing such terminals should be decided by market forces" (Tilley 1989: 33). By the early 1990s, the French system too had succumbed to commercial pressures; charges of £80 a terminal were introduced in order to recoup some of the costs.

The development of fully integrated EHS systems selling both convenience and comparison goods would enhance consumer access to most shopping activities. This should be seen as a positive way of maximizing shopping opportunities for the public. This is arguably the most important issue of new technology in retailing, as awareness of the relatively high numbers of "disadvantaged" and "neglected" consumers throughout the UK has increased. The advent of micro-technology also has the potential to revolutionize shopping activity. The provision of telecommunication links between a superstore and focal points in a local community has meant that deprived consumers, even with limited mobility, have been able to place orders and receive deliveries at their homes. The more affluent have been able to order from home and have the goods delivered there.

For consumers as a whole, the increase in access to shopping facilities will depend on the cost of installing systems and the services those systems provide. If large integrated systems were to be set up, local shops might be used as drop-off points for goods, thus allowing planners to support the retention of local shopping facilities (Guy 1985). Any increase in consumer access to shopping facilities, especially if this were to involve the whole population, should be viewed positively. It has to be recognized, however, that the development of EHS could have a detrimental impact on existing shopping facilities. The closure of any stores due to the introduction of EHS could have an adverse impact on consumer access.

The development of EHS could potentially give considerable advantages to all consumers, mainly in terms of time and convenience, but also by giving them wider access to goods. Marti & Zeilinger (1982), and Foster (1981) identified the advantages of EHS to consumers, which are summarized by McKay & Fletcher (1988: 2), as:

- the elimination of drudgery associated with queuing up to pay for goods;
- quickness and convenience;
- the elimination of carrying goods home by door-to-door delivery;
- easy comparison of prices;
- breadth of choice;
- the relief of high street congestion;
- the relief of transport/parking problems;
- the elimination of transport costs; and
- increases in leisure time.

Consumer advantages are going to depend on a variety of factors associated with the development of EHS. First, there is the cost of electronic equipment. If equipment is distributed at low cost in large numbers, then a "critical mass" of users could be achieved quickly. This could in turn attract a variety of information providers. The question is who will provide such a terminal? If electronic equipment is to be bought, it is probable that the equipment will only be acquired initially by the higher socioeconomic groups, because the financial outlay is likely to be relatively high. Such a policy has failed for Prestel and would seem unlikely to work for any other system. If electronic equipment were given away, through subsidies from public agencies in an attempt to increase access to shopping facilities for the disadvantaged, those who received such facilities would derive considerable advantages. Davies (1985) argues that such schemes as those found in Gateshead and Bradford experience great difficulty in expanding to more than 500 users, because of increased costs. This would mean that only a small minority of disadvantaged consumers would actually receive the benefits from any future social EHS schemes. Consumers, as a whole, would derive the largest benefits if the electronic equipment required were given away free and a combination of commercial and subsidized schemes were developed. Guy (1985) argues that major firms could be encouraged to set up such schemes, perhaps through Section 106 of the Town and Country Planning Act 1990 (Telling 1990), with agreements attached to planning approvals for new retail development.

Secondly, advantages will depend on the types of services available to users. Prestel gives an indication of both the potential and the limitations of this new technology. This system is ideal for providing information about products and services, but information-providers are unlikely to accrue significant revenues unless they receive an income, either through the sale of goods or services, or the sale of space utilized. On Prestel a majority of the items for sale do not require visual inspection, for example, tickets; or they are very standardized, for example, wines. Consumers are only likely to be attracted to use EHS services if they receive benefit through convenience, availability, or prices offered, or perhaps a combination of these factors. To maximize benefits for consumer participation retailers will need to consider what services to provide. Users are likely to be attracted to specific facilities.

Thirdly, consumers will be attracted to the speed and efficiency of EHS, but if retailers enter on an experimental basis they may not be able to maximize efficiency and therefore consumer convenience may not be very high and services may fail. Those consumers already involved in mail-order shopping would increase their speed of ordering, and as a result might receive goods more quickly, although any savings on postage might be outweighed by equipment and operating costs.

Various disadvantages of EHS to consumers have also been identified by McKay and Fletcher (1988) and in *Which*? (1991: 149):

- unauthorized use of a terminal may occur;
- risk of error in ordering is high;
- high-pressure sales methods may be used;
- ambiguous information may appear on the screen;
- long delays in receiving goods ordered and paid for may occur;
- consumers may change their minds about their orders;
- goods may not be delivered;
- goods differing from those the consumer thought they would receive may arrive; and
- without a printed record it may be difficult to establish which party is responsible for any error.

The consumer impact of EHS depends on the types of systems and services information-providers deliver. In the short term, only limited numbers of consumers are likely to have access to such facilities. If they can afford to pay for such facilities and are willing to use them, they can benefit from increased shopping opportunities. It must be accepted that many people enjoy shopping, and often those who are elderly and mobile use this activity as a way of getting out of their homes and meeting people. Such systems and services, in the beginning, will be used mainly by those wishing to maximize leisure time and who can afford to pay for such services. This may do nothing to help those who have limited access to shopping facilities; rather it may increase access for the most affluent consumers.

EHS has the potential to increase access to all shopping facilities for the disadvantaged. If subsidized, EHS would allow the elderly, disabled and housebound to make orders from home and have goods delivered there. It is hoped that in the long term commercial EHS schemes will be established to run alongside subsidized systems, and further maximize consumer access to all shopping facilities. Therefore the problems faced by both "disadvantaged" and "neglected" consumers could be tackled through further public-sector-aided provision of electronic shopping, in schemes such as those already tried in Gateshead, Bradford and the London Docklands. It has been suggested that major grocery firms may be encouraged to set up such schemes through Section 106 agreements attached to planning approval for new superstores or, eventually, automated warehouses (Guy 1985). Such subsidies may be a solution for the small group of "neglected" consumers, but Davies & Champion (1980) conclude that more than 25 per cent of the population can be described as disadvantaged in terms of access to shopping provision. The enormous cost of providing a system for both groups of deprived consumers means that schemes are likely to be available only to the relatively small number of "neglected" consumers, those who are housebound, disabled or elderly and have limited access to shopping facilities. The much larger number of disadvantaged consumers would be forced to purchase terminals and pay for delivery of goods, as retail firms or the government would be unlikely to provide large subsidies to this group. If these disadvantaged consumers are forced to pay for terminals and delivery they are unlikely to be any better off than they are at present.

EHS is very much in its infancy in the UK and it is probable that experimental schemes will continue to flourish. However it seems unlikely, especially in a fairly depressed economic climate, that large-scale systems selling a wide variety of services will emerge from this experimental stage in the near future. Nevertheless, the speed of development and introduction of new forms of information technology in the 1980s would suggest that developments in EHS could take place very rapidly if a suitable terminal were developed, with financial backing from a large number of major retail firms. It is hoped that if large-scale EHS services are developed they will be utilized positively for the population as a whole, but particularly for the needs of disadvantaged consumers.

Conclusions

Changes in retailing have meant that some people do not have equal access to shopping facilities. These are the disadvantaged consumers, who are not satisfactorily catered for by the retail hierarchy as it currently operates. Research work in this area is limited, but it is clear that many groups with individual retailing needs are not satisfied. Solutions to consumer deprivation have been partial and really only address some issues for individual groups, for example, introducing improved access facilities for the disabled. One of the great difficulties in tackling the problems of disadvantaged consumers is that the public sector has only limited power to intervene in the marketplace. As a result, consumers depend heavily on private-sector policies, which are geared towards profit maximization. Current developments within the private sector look likely to help the disadvantaged, with the movement of limited-line discounters into the UK. Firms such as Aldi have tended to locate in low-income neighbourhoods and to sell competitive convenience goods.

The issue that unites all disadvantaged consumers is the problem of access to the larger and cheaper retail outlets. The aim should be to maximize consumers' "potential mobility", and therefore allow them to shop at the greatest variety of facilities. This can be achieved in two ways: first, by a store location policy, which for example encourages stores to locate in areas easily accessible to disadvantaged consumers. This can be done through the planning system, but requires both strong retail planning-policy guidance from central government and research by local authorities to identify suitable areas for store location. Local authorities should attempt to develop much closer links with retail firms. Secondly, "potential mobility" can be considerably enhanced for those without cars and those who cannot drive by the provision of efficient and cheap public transport. Public transport is currently expensive and inefficient, and it is difficult to use for convenience shopping, because of the amount of goods to be carried and stored.

EHS is a potential new solution to the problems of disadvantaged consumers. If terminals were located in the home and goods delivered at a competitive price, consumer access to the widest variety of goods could increase. It is accepted that individuals will need to be connected into systems to participate, but experience in France has shown that this is possible. The larger question is whether retailers would be willing to subsidize disadvantaged consumers? Moreover, the social function of retailing should not be forgotten in any debate about EHS. Many individuals rely on daily shopping trips as their only contact with the outside world, therefore the introduction of EHS for the disadvantaged would need to be carefully managed. Planning for disadvantaged groups can benefit the community as a whole, since initiatives may "provide indirect savings to the taxpayer by enabling disabled and elderly people to live independent lives" (Bowlby 1985: 220).

CHAPTER TEN
Working conditions and the trading week

David A. Kirby

Compared with other aspects of retailing and retail management, employment is under-researched. As Sparks (1987: 239) has observed "employment in retailing has, historically, been one of the least-considered aspects of the distributive trades". With the increasing awareness of the importance of services in general, and retailing in particular, to the UK economy, this situation has begun to change. Since the early work of Robinson & Wallace (1976) on pay and employment, significant studies have been undertaken of part-time working in retailing (NEDO 1988b), the employment characteristics of large stores (Sparks 1983) and, more recently, the training needs of the retail workforce (Robinson 1990, Jarvis & Prais 1989). With the population of working age set to experience profound structural changes, and the civilian labour force in the UK, as elsewhere, set to have a more diverse composition than it does at present, employment seems likely to emerge as a most important issue in contemporary retailing.

For some time, two apparently conflicting trends have been emerging in Western labour markets: the tendency towards longer business operating times and shorter personal working hours. The argument for the former is that it leads to increased output and economic growth potential and, when coupled with shorter working hours, increased employment and prosperity levels. For such a strategy to succeed, a labour force that is more flexible and diverse than has been traditional is required. Given the sharp fall in the number of school-leavers, various groups from the adult population are expected to form a growing part of the workforce. These, it is predicted, will include those over 55, ethnic minorities, those currently unemployed, women and the disabled. Already some UK retail companies are beginning to formulate policies to extend the labour force by attracting such previously under-represented groups, most notably women, and by extending the hours of working through shift and weekend working.

This situation is of particular significance in retailing. Women already constitute about 54 per cent of the retail labour force and, as Robinson (1990: 284) has observed, there is also "a large and growing part-time component (some 35 per cent rising to almost 50 per cent)" in the 2 million employees in retailing, and increasing pressure to extend trading hours. Essentially these latter are controlled by the Shops Act 1950, which is a consolidating act reproducing the provisions of the Shops (Hours of Closing) Act 1928, the Shops (Sunday Trading Restriction) Act 1936, and the Retail Meat Dealers Shops (Sunday Closing) Act 1936. The legislation, which had its origins in the late 19th-century concern to protect shop workers from exploitation, relates to conditions very different from those that now exist. The legacy is anachronistic trading hours that are difficult to justify, particularly since the legislation contains inconsistencies and injustices relating to the types of products that can be sold on the Sabbath. In summary, the legislation states that:

(a) Shops must close by 8pm on a weekday and 9pm on Saturday or one other day laid down by the local authority;
(b) Shops must close one afternoon a week by 1pm. Traders may choose their own closing day, although the rule may be relaxed if the majority of traders in an area agree;
(c) A shop may not open on Sunday, except for the sale of:
    - intoxicating liquors
    - meals and refreshments to eat on or off the premises, but not fish and chips at a fish-and-chip shop
    - newly cooked provisions and cooked, or partly cooked, tripe
    - table waters, sweets, chocolates, sugar confectionery and ice cream including wafers and edible containers
    - flowers, fruit and vegetables, including mushrooms, but not tinned or bottled fruit or vegetables
    - milk and cream (not tinned or dried milk or cream) and clotted cream sold in tins or otherwise
    - medicines and medical surgical appliances
    - aircraft, motor or cycle supplies or accessories, but not cars,
    - tobacco and smokers' requisites
    - newspapers, periodicals and magazines
    - books and stationery at bookstalls on railway stations and approved airports

- guidebooks, postcards, photographs and souvenirs at any gallery, museum, garden park or ancient monument, zoo or in any ship carrying passengers
- passport photographs
- sports equipment at a place or premises where that sport is played
- fodder for horses, mules, ponies and donkeys at any farm, stables, hotel or inn
- post offices and undertakers' businesses.

Also, local authorities can, if they so wish, make an exemption order permitting the sale of bread, fish and groceries until 10am on a Sunday, and in holiday resorts shops are permitted to be open for not more than 18 Sundays in the year to sell:

- bathing equipment or fishing tackle
- photographic requisites
- toys, souvenirs and fancy goods
- books, stationery, photographs, reproductions and postcards
- any article of food.

Apart from these permitted exceptions, and those for Jewish traders who close on Saturdays, no shop may be open on a Sunday.

The legislation is enforced by local councils, which vary in their attitudes to it. The result is that "enforcement shows marked variations between different parts of the country" (Wrigley 1984: 223), and increasingly the law is being flouted. There would, therefore, seem to be a *prima facie* case for reform, to ensure that, in the words of former Prime Minister Margaret Thatcher, the hours of retail trading "correspond to modern patterns of living".

Much has been written about the implications of any revision of the law. In particular, attention has focused on the implications for the structure of the retail system (Kay et al. 1984) and the likely economic implications for retailing of extended trading hours (Burke 1986, Dawson 1985). By contrast, relatively little attention has been paid to the implications for those for whom the legislation was designed largely to protect: retail employees. This chapter will consider some of the direct and indirect implications for employment planning in retailing resulting from any attempt to extend significantly the hours of retail trading.

Trends in the workforce

Since the most recent attempt to revise the law in 1986, the question of employment and the condition of the labour market has become an issue of increasing significance, not just in the UK but throughout western Europe and the rest of the developed world. As mentioned above, this has resulted essentially from the demographic changes occurring in society, resulting from a decline in the younger age cohorts, and the consequent need to expand the workforce in order to maintain growth and productivity. This is particularly important at a time when the tendency is towards shorter personal working hours but longer plant operating times. Thus the context for the chapter is not simply the changing organization of labour within British retailing, but the way employment is organized generally and, with the completion of the European Community's single market, particularly in Europe.

Here, according to a recent EC survey (Commission of the European Communities 1991) the female share of total employment rose in every member state, largely because of the growth of part-time jobs in the service sector. Women now account for 38 per cent of total employment in the EC. Currently, about 15 per cent of employees in the EC are part-time, though in the UK the proportion reaches 21 per cent. The majority of the EC's part-time workers are women, and one-third of women working full-time would prefer to be part-time. Shift work is undertaken regularly by about 16 per cent of all employees in the EC, being most common in the UK and Spain. Some 61 per cent of all employees would be willing to work unsociable hours (early or late shifts) and 44 per cent expressed a willingness to work on Saturdays. Approximately 18 per cent of European employees work on Sundays, but in Greece, the Netherlands and Portugal employees would like to see less Sunday work. However, in the UK (where the proportion of Sunday working is above the EC average), and in Denmark, France and Ireland, it would appear that noticeably more workers are prepared to work on Sundays than do so already. Even so, some 75 per cent of European workers (61 per cent in the UK) seem to be unwilling to work on Sundays, even if offered higher wages or additional leisure time (Commission of the European Communities 1991).

Within this European context, the trend in British retailing has been for increased feminization of the workforce and for the number of part-time jobs to grow, often at the expense of full-time employment. While the reasons for this have been outlined elsewhere (Robinson 1990, Sparks 1987), it is difficult to trace with any precision the changes that have occurred because of differences in the methods of data collection and classification in the main official statistical series (Reynolds 1983). However, on the basis of the evidence available, it would seem that the total retail workforce has declined from 2.9 million in 1961 to 2.1 million in 1992 (Employment Gazette 1992). The proportion of female employees has increased from 61 per cent in 1961 to 63 per cent in 1992 and the proportion of part-time jobs from 28 per cent to 47 per cent (Employment Gazette 1992). Associated with such trends has been the deskilling of the workforce and an increase in the complexity and diversity of patterns and conditions of employment. As a consequence, British retailing "offers a 'bundle' of employment opportunities" (Sparks 1987: 250) and there is "considerable and increasing diversity in the labour packages which make up an organization's human resources" (Robinson 1990: 301).

It is against this background that the chapter considers the attitudes of retail employees in the UK to working non-standard hours, Saturdays and Sundays, and attempts to consider some of the possible employment implications resulting from any change in the law regulating retail trading hours.

A questionnaire survey of retail employees

To meet these objectives, a questionnaire survey of retail employees was undertaken in August 1991. The survey was conducted on a self-completion basis, the questionnaires being left with the respondents and, where possible, collected by the research team on completion so as to minimize possible management influence and ensure confidentiality and impartiality. In some instances, respondents requested more time to complete the questionnaire and these were mailed back.

In total, six retail organizations agreed to participate in the survey, providing a total sample size of 1,179 employees in 41 stores. Out of this, 483 completed, usable questionnaires were received, yielding a 41 per cent response rate from 37 outlets. The resultant sample represents a range of trade sectors, including bakers, department stores, and DIY, footwear, grocery and hardware outlets. The stores employed between 9 and 119 employees and were spread throughout England and Wales, and also Scotland, where Sunday trading is permitted by law. The sample contains responses from employees who work on Sundays, because one company, providing 30 per cent of the respondents, trades on Sundays. In total, 18 per cent of the sample are required to work frequently on Sundays. While the national proportion of employees in retailing required to work on Sundays is not known, some 22 per cent of the UK workforce does so. Thus it would appear that the sample is not biased against Sunday workers.

The questionnaire was divided into five sections, focusing on the respondent, the job, the respondent's domestic circumstances, attitudes to work and leisure activities. It took approximately 15 minutes to complete and was tested before full implementation. Because of confidentiality assurances given to both respondents and participating companies, the responses were not recorded by either company or locality. Thus analysis of the resultant data is possible at the aggregate level only.

Research findings from the survey

Profile characteristics of the sample

A total of 483 employees, all women, participated in the survey, with the modal age group being 21–30 years (Table 10.1) and approximately half of the sample aged between 21 and 39. An overwhelming majority of the respondents (92 per cent) were in non-managerial positions, confirming earlier findings on the de-skilling of the retail labour force (Robinson & Wallace 1976) and its subsequent polarization into a small band of specialist managers at one end of the spectrum and a large "army of labourers" at the other (Alexander & Dawson 1979). Some 50 per cent of the sample possessed no formal academic or professional qualifications and few had been educated beyond "O" level or its equivalent. In approximately 60 per cent of the cases, slightly higher than might be expected, work was part-time, although respondents' ages would appear to influence work categories. Table 10.2 indicates that most of the women aged between 21 and 25 were full-time workers (72 per cent), whereas the proportion of women working full-time drops dramatically to 39 per cent and 33 per cent, respectively, in the age groups 26–30 years and 31–39 years.

Of the total sample, 56 per cent stated that they had children, though of these, half had children who were already working or were over the age of 18. Seventeen per cent had children of pre-school age, while 29

Table 10.1 Age categories of respondents.

Age group	%
20 years and under	13.9
21-30 years	30.5
31-39 years	19.3
40-49 years	21.4
50 years and over	14.9

Table 10.2 Ages of women and their work status (%).

Age band	Full time	Part time
20 or under	48.5	47.0
21–25 years	72.1	24.4
26–30 years	39.0	59.3
31–39 years	32.6	67.4
All age groups	40.8	56.6

Table 10.3 Age of respondent compared with ages of children (%).

	Ages of Children		
Age band	Preschool	Primary	Secondary
21–25 years	18.2	1.3	–
26–30 years	54.5	29.3	7.1
31–39 years	25.0	56.0	41.4

Table 10.4 Child-care provision during working hours (%).

	Hours of work			
Carer	Normal	Non-standard	Saturdays	Sundays
Partner	16.8	45.7	51.1	60.8
Relative	49.5	35.7	30.9	22.8
Partner & relative	14.7	12.9	12.8	23.9
Child minder	7.4	2.9	3.2	1.3
Nursery	2.1	1.4	-	-
Other*	9.5	1.4	2.1	1.3

* Including the children being at school or being cared for by friends or neighbours.

per cent and 27 per cent, respectively, had children at primary school and secondary school. For those with dependant children (Table 10.3), both the women's ages and those of their children would appear to influence patterns of participation in the labour market, thereby affirming the findings of Martin & Roberts (1984). Of those women with children of pre-school or school age, the majority relied upon their partners and/or members of their families for child care (Table 10.4), suggesting there is either a lack of appropriate child-care provision or that the cost of private child care is generally prohibitive. As one respondent commented, "if I wanted to work full-time, I would have to pay a child minder more than I earned".

In addition to caring for children, 12 per cent of the sample indicated that they cared for other dependants (such as elderly or handicapped relatives). The frequency of care varied, with some 7 per cent of this sub-sample caring for dependants once a month and 35 per cent once a week. However, for 42 per cent of the group the level of care required entailed daily or more frequent visits. Of the remaining 15 per cent, most commented that their dependants, primarily elderly parents, lived with them and so care was effectively 24 hours a day. Thus, care of relatives is a significant responsibility for a minority of the sample, and the trend for women to become carers in middle age, noted by the 1989 CBI *Survey of women and work*, is affirmed by the findings. Whereas fewer than 7 per cent of the respondents were carers under 30 years old, the proportion was 22 per cent for those aged 50 years or over.

As Table 10.5 indicates, approximately half of the sample travel to work by car; this suggest that if required to work non-standard hours and at weekends, the journey to work would not be a problem. However for a significant minority, transport to and from work outside the normal working day would present difficulties. While only 9 per cent of

Table 10.5 Mode of transport used to get to work (%).

Transport mode	Hours of work			
	Normal	Non-standard	Saturdays	Sundays
Bus	34.0	26.5	30.7	15.9
Car	41.3	51.5	46.0	58.6
Train	1.6	0.8	1.7	0.9
Walk	16.2	17.0	17.0	19.4
Other*	6.8	4.2	4.6	5.2

* Includes bicycle, moped and lifts from friends or colleagues.

those working on Saturdays stated that they experienced transport problems, the proportions were 17 per cent for non-standard hours and as high as 30 per cent for Sundays.

The job and attitudes to work

With three-quarters of the sample (76 per cent) earning less than £6,239 per annum (£120 per week), levels of remuneration are noticeably low (Table 10.6). Given the predominance of part-time working and non-managerial grades, this is explicable but, nonetheless, the findings confirm those of earlier studies, most notably the work of Robinson & Wallace (1976), Pond (1977), and Hurstfield (1978), and point to the long-term dependence of British retailing on low-paid female labour.

Table 10.6 Salary and work category of sample population.

	% of women who work	
Salary (per annum)	Full time	Part time
Under £2,079	0.4	12.1
£2,080–3,639	0.6	31.2
£3,640–6239	17.2	13.0
£6,240–9359	18.5	0.6
£9,360–15,599	3.8	0.4
Over £16,000	0.4	–

However, only a small minority of the respondents (11 per cent) are solely dependent on their job for income, although over a quarter (27 per cent) work to provide essential income for their households, and income-related factors are clearly of significant importance for the majority of respondents (Table 10.7).

Some 60 per cent of the sample were occupied on a part-time basis and from Table 10.8 it is clear that the modal class of hours worked among the sample is 16–29 hours a week, reflecting the part-time nature of retail employment. However, Table 10.9 reveals that approximately 70 per cent of the respondents are required to work regularly on Saturdays, 35 per cent non-standard hours and 18 per cent on Sundays.

Table 10.7 Respondents' reasons for working.

Reasons	%
Sole breadwinner	10.8
Essential income for household	27.3
To supplement household income	41.9
Want financial independence	34.4
To meet people	30.2
For career development	17.3
For interest	24.4
Other*	2.9

* Including money for luxuries and "extras" as well as for hobbies.

Table 10.8 Hours of work per week.

Hours	%
Less than 16 hours	19.9
16 to 29 hours	34.9
30 to 39 hours	32.2
40 to 49 hours	11.8
50 hours or more	1.2

Table 10.9 Women currently required to work unsocial hours (%).

Category	Non-standard	Saturdays	Sundays
Always	18.3	37.9	6.0
Regularly	16.9	30.2	12.0
Seldom	23.3	15.2	18.9
Never	41.5	16.7	62.8

Attitudes towards working non-standard hours, Saturdays and Sundays

When asked about preferred working patterns, 43 per cent of the respondents said that they would want to continue working their current hours (Table 10.10), with 40 per cent preferring not to work on Sundays and 30 per cent on Saturdays.

Table 10.11 analyzes, further, the preparedness of the respondents to work non-standard hours, Saturdays and Sundays. It reveals that in the case of non-standard hours, 64 per cent of the sample would be prepared to work them, though only a small minority (8 per cent) would be prepared to work them "all of the time". By contrast, Saturday working would appear to be more attractive to the respondents than working either non-standard hours or Sundays. Almost 20 per cent of the sample stated that they would be prepared to work every Saturday, with a further 30 per cent prepared to work regularly on Saturdays. Only 15 per cent did not want Saturday work.

Table 10.10 Preferred patterns of work.

Preferred pattern	%
Start work earlier	13.3
Finish work earlier	21.1
Have more flexible hours of work	15.4
Work extra hours for more time off	7.6
Work a regular 9 to 5	19.4
Do no Saturday work	29.9
Do no Sunday work	40.0
Work days not nights	15.6
Work within school hours	9.9
Work current hours	42.5

Table 10.11 Women prepared to work unsocial hours (%).

Attitude to working	Non-standard	Saturdays	Sundays
Always	7.9	19.1	6.6
Regularly	20.7	30.0	13.8
Seldom	35.6	23.8	20.6
Never	25.1	14.6	50.8
Not applicable	10.8	12.5	8.2

Whereas 15 per cent of the sample would not work on Saturdays and 25 per cent were not prepared to work non-standard hours, 51 per cent were not prepared to work any Sunday hours and a further 21 per cent only rarely. When asked what they would do if their employers required them to work on Sundays, some 42 per cent of the sample claimed that they would try to change their jobs, while a further 31 per cent said they would be unhappy about working on Sundays but would have to stay in their jobs. Only 28 per cent of the sample would be happy to stay in their jobs if required to work on Sundays, compared with 44 per cent for non-standard hours and 65 per cent for Saturday working. For those prepared to work non-standard hours, Saturdays and Sundays, the most significant inducement is clearly premium payments (Table 10.12), particularly for work on Sundays, when double-time payments are required (Table 10.13).

Table 10.12 Conditions necessary for respondents to work non-standard hours, Saturdays or Sundays (%).

Conditions necessary	Non-standard	Saturdays	Sundays
Creche facilities	7.2	5.8	4.6
Transport	10.2	5.3	12.3
Flexible hours of work	21.0	23.3	12.8
Time off in lieu	20.0	25.3	17.0
Premium payments	39.5	38.0	50.2

Table 10.13 Premium payments required for the working of unsocial hours (%).

Payments required	Non-standard	Saturdays	Sundays
1.25 times normal	9.5	4.6	1.2
1.5 times normal	21.3	21.1	6.1
Double time	24.1	13.1	52.9

Hours of work and domestic life

Some three-quarters of the sample have partners and Saturdays (57 per cent) and Sundays (72 per cent) are the only days when most of them are able to be with them. For the majority of the sample, the time they currently have for their families was felt to be satisfactory but when questioned about the likely impact of working non-standard hours, Saturdays and Sundays, fewer than 10 per cent thought that it would improve the situation. For 60 per cent of the sample, Sunday working would reduce time with their families, whereas only 40 per cent believed that this would be the case with Saturday working. Similarly, only a small minority of respondents believed that working unsocial hours would benefit their relationships with their children, partners and families, and for a significant proportion such hours were believed to be potentially harmful (Table 10.14).

Table 10.14 The perceived impact of working unsocial hours upon the respondents' relationships with their partners, children or families (%).

Perceived impact	Children	Partner	Family
Benefit	5.7	5.5	5.8
Harm	39.9	42.2	35.8
Neither benefit nor harm	41.7	37.9	44.3
Don't know	12.7	14.4	14.3

The responses to other questions in the survey reveal that only about 16 per cent of the sample are currently dissatisfied with the amount of time they have available for leisure activities. Just over half of respondents are satisfied with their amount of leisure time. The array of leisure activities pursued is broad-based but focused strongly on the family, including relatives and friends, especially on Sundays (Table 10.15). Therefore if there were to be an increase in the frequency of working non-standard hours, Saturdays and Sundays, the time respondents have available for family and friends, as well as other leisure pursuits, would be likely to be affected considerably.

Table 10.15 Leisure activities (%).

Activity	Weekday	Saturday	Sunday	Days off
Watch television	60.7	42.2	45.5	41.0
Time with family	38.5	39.1	56.5	42.7
Play sport	14.4	5.8	7.1	9.6
Visit friends	28.6	23.4	27.7	24.1
Visit relatives	19.6	21.5	41.7	28.3
Go to church	1.0	0.6	10.7	0.6
Gardening & DIY	15.0	15.1	24.1	23.1
Go shopping	47.2	39.1	7.1	52.6
Other*	10.6	10.1	10.2	12.1

* Including housework, trips to the theatre and cinema, day trips etc.

Conclusions

At a time when female "returners" are set to become an increasingly significant element in the UK workforce, British retailing remains highly dependent on part-time female labour. At the same time, working conditions appear not to be as attractive as they might be. Pay remains low and career prospects and training opportunities are poor, especially for part-time employees, who make up the vast majority of the retail workforce (Robinson 1990). Traditionally, one of the main attractions of employment in retailing, particularly for the low-qualified woman with family commitments, has been the flexibility of working hours. As Sparks (1987: 250) has recognized, many "employees of the required quality may . . . have almost 'tailor-made' hours of work". With the proposed liberalization in retail trading hours, this may no longer be the case. As elsewhere (Price & Yandle 1987), the pressure to change the law on the hours of retail trading has resulted from the broad changes that are occurring not only in the behaviour of consumers but in society and the way work is organized. Conversely, however, any change in the law is likely not only to have an impact on the operation of the retail system but also to have wider implications for society in general. In particular, complete liberalization of the hours of retail trading, as recommended by the 1984 Committee of Inquiry into Proposals to amend the Shops Act (Home Office 1984), would have considerable implications for the lives of shop workers, the very group it was introduced to protect.

Increasingly, it would seem, consumers seek to shop on Sundays. However, while a Mori survey in November 1991 discovered that 70 per cent of consumers said that they would shop on Sundays if stores were open, it also found that 90 per cent of them believed that shop workers should have the legal right to refuse to work on Sundays. As has been demonstrated, a large proportion of retail employees, particularly those with children, are reluctant to work unsociable hours and especially on Sundays. At the same time, however, many of them are vulnerable in the employment market: they have few formal qualifications and are heavily dependent on the income from their jobs. Under such circumstances, the large proportion of those not happy to work unsociable hours, and particularly Sundays, might have difficulty finding alternative employment and would have no alternative but to agree to Sunday work.

One solution might seem to be strong employee protection. While such a solution might find support in the Social Charter of the EC, the main problem with any such legislative solution would be policing, as employees, because of their vulnerability, might be particularly reluctant to bring cases against their employers. Additionally, it is notoriously difficult to document informal and subtle pressures placed on employees by employers. As one retail employee from the northeast of England is reported to have commented when her employers opened their store for Sunday trading: "It's supposed to be voluntary but it's expected you'll turn up" (Robertson 1991).

Assuming legislation were introduced and was effective, the implications for retailing would be considerable. From the evidence provided here, it would seem that the proportion of employees wanting neither to change their existing hours of working nor to work on Sundays would be so large as to make if difficult for retailers wishing to extend their hours to staff their stores from their existing staff complements. This would imply that such operators, in the short term at least, would be required to have a specialist cohort of staff for unsocial hour working. Clearly the costs involved in such an arrangement could be considerable, particularly for those businesses where knowledge of the product and customer preference is critical and staff training and development important. Additionally, it must be borne in mind that as the demographic downturn in the labour market takes hold, competition for staff, particularly women returners, will intensify and remuneration packages will have to become more attractive. Apart from including sales-related bonuses, it would seem that such packages would have to include premiums for working unsocial hours, allowances for care assistance for both children and more elderly dependants, and, possibly, assistance with the journey to and from work. Under such circumstances, the profitability of extended hours of trading could be affected considerably, particularly if, as Ferris (1991) suggests, shopping hours tend to be extended in response to the demands of the marginal customer.

Finally, it needs to be recognized that in most high-level economies, and throughout Europe, the trend is towards a shorter statutory working week and increased time for leisure activities. The UK is not different in this respect and there is a clear case for harmonization of UK employment legislation on the hours of retail trading with those of its EC partners. Against this, shopping is increasingly a leisure activity and consumers should be able to shop at times convenient to themselves. With the trend towards increased female participation in the workforce, this suggests that retail outlets should be open outside the normal working week. However, the hours of work, generally, are being reduced and more time is becoming available for shopping during non-standard hours (after 5 pm and before 9 am) and on Saturdays. As both times tend to be preferred by women shop workers to Sunday working, those retailers wishing to expand their retail provision might prefer to explore the opportunities afforded by these solutions, particularly as government restrictions on the development of additional provision out-of-town, which were in place in 1984, have subsequently been removed.

While the consumer may legitimately argue for longer opening hours, policy-makers responsible for planning the availability of retail provision must be careful to weigh this against the welfare needs of those engaged in retailing - a group that remains vulnerable in the labour market and which the original shop legislation was designed to protect.

Acknowledgements

The research on which this chapter is based was funded by the Keep Sunday Special Campaign and conducted by Paul Livett and Dawn Rotheram. The views expressed are not necessarily those of either the sponsor or the research team.

CHAPTER ELEVEN

Shopping as leisure

Peter Newby

In a society built on the mass consumption of goods and services it might seem strange to raise questions about the position of shopping in our lives. It is important, however, for the continued vitality and profitability of the retail sector that these issues are explored. The heart of the argument in this chapter is that, while people enjoy spending money, their enjoyment of consumption is frequently frustrated by a retail system that strips the "pleasure" element from shopping and reduces it to a mere economic exchange of goods for money. It is apparent that so far this approach has benefited the retail sector, but in the longer term those retailers and developers who do not appreciate how current retail provision can give rise to consumer frustration will suffer.

This chapter explores a number of issues of crucial importance to the retail sector. First, it is important to recognize that there is not just a single category called "shopping", and that to lump all purchase behaviour together will obscure the significance of differences in shopper motivation for retail opportunity. The second issue for the retail sector to address is the need to recognize and respond appropriately to the emergence of shopping as a leisure activity. While the meaning of leisure is explored in this chapter, it is clear that this market segment is not well understood and that the lack of empirical research inhibits the response that the sector might make. The third issue concerns the problems facing retailers who try to square their drive for efficiency and market share with the needs of a growing leisure shopping sector. In particular, retailers and centre managers have to find ways of confronting shopper boredom (which arises from predictability in the shopping environment) so that it does not undermine the leisure dimensions of shopping. The chapter concludes by considering how the sector has responded and could respond to the opportunities that such shopping offers.

The growth in leisure

It is clear that reductions in working hours have led not only to increased leisure time but also to further economic opportunities. This process began in earnest in the UK in the 19th century with a sequence of legislation (the Factory Acts) that regulated working practices and working hours; for factory workers, the 1833 Act established a 10-hour working day and the 1850 Act established Saturday as a half day (Cunningham 1980). The 1874 Act reduced the standard working week from 60 to 56½ hours. Other legislation consolidated Christmas Day and Good Friday as holidays. Frequently, the legislation was a response to pressure groups, for example, the 1850 Act resulted from the pressure exerted by the Early Closing Association. As Cunningham (1980) indicates, however, as important as the legislation was the power of organized labour, and within a few years of the 1874 Act many unionized groups were working nine-hour days. From this point on it was labour organization rather than legislation that reduced the working week and increased the time available for recreation. However, while it is undoubtedly correct to see the balance between work time and leisure time shifting over the past two centuries it would be incorrect to assume that workers of a century ago took little time off. Cunningham (1980) notes that evidence presented to the 1876 Royal Commission on the working of the Factory Acts suggests that some workers took 20 days holiday a year in days and half days and it was so common among workers in many sectors to take the Monday after payday off that it became known as "Good Monday".

However, increased affluence has been as important as the reduction in the working week for the growth in leisure. This has led to the commercialization of leisure, including everything from books and gardening to holidays, cinema and now, as demonstrated in this chapter, shopping.

The use of leisure time

Popular use of the term "leisure" suggests that it is an active process frequently requiring planning, organization and expenditure. Respondents to the General Household Survey, for example, will freely mention sports, visits and entertainment but often require prompting for the most common leisure activities, such as watching TV, social visiting, and listening to radio, records and tapes.

The most popular leisure activities of visiting friends, watching TV and listening to the radio, are engaged in by the great majority of the adult (over 16) population (Table 11.1). The proportions were remarkably stable in the 10 years 1977–86. The same is true of other activities whose main characteristic is that they are enjoyed at home and have the object of using up time. These activities are more important than active participation in all sports and physical activities, watching all sports, going on outings or sightseeing or making use of all entertainment facilities. What is more, with the exception of active participation in sports and sightseeing, the proportion of people engaged in active leisure has fallen, while the proportion engaged in passive leisure has risen or remained more or less the same. Quite clearly there are shifts taking place in the way we spend our non-working time.

Table 11.1 Adult population engaging in specified leisure activities (%).

(a) Unprompted mentions of leisure activities

Year	Sports Active	Sports Spectator	Outings	Sight-seeing	Dancing	Cinema
1977	21	10	16	13	15	10
1986	28	8	14	15	11	8

(b) Prompted mentions of leisure activities.

Year	Visiting friends	TV	Listening to Radio	Listening to Record/Cassette	Reading	Gardening	DIY
1977	91	97	87	62	54	42	35
1986	94	98	86	67	59	43	39

Source: OPCS 1987.

As Gershuny (1987) has shown, the shift from work time to leisure time has continued. This creates more leisure time when adult involvement in active leisure opportunities is generally declining or stable. He demonstrates that between 1961 and 1985 "three quarters of an hour, in the average day of the average adult . . . moved out of 'the realm of necessity'" (Gershuny 1987: 494). A significant proportion of this, up to 45 minutes according to Gershuny's calculations, came through the application of labour-saving devices to domestic labour. Three-quarters of an hour may not sound a lot, but the 30 minutes of this that was spent in consuming services outside the home represented an increase of 50 per cent between 1961 and 1985. Those 30 minutes produced 1.7 million additional jobs in the service sector. Over the same period, Gershuny shows that the average person spent an extra 35 minutes a day on other domestic tasks, including shopping, amounting to an almost 100 per cent increase over the 15-year period.

The reasons for this extra time spent shopping may, as Gershuny suggests, reflect the process of shopping in large supermarkets. He puts forward the view that larger supermarkets mean more time walking up and down the aisles, and probably more time queuing to pay for goods (Gershuny 1987: 497). There may be an element of this, but the effect is almost certainly outweighed by the time saved in queuing several times for counter service at one shop before moving on to another, and by a reduction in the frequency of food shopping as the supermarket became established on the British retail scene.

If this analysis is correct, these figures have implications for shopping behaviour over the period. They imply that the increase of 35 minutes per day underestimates the increase in time spent shopping for non-food and household goods. In other words, shoppers are spending more time in stores selling non-household goods. Since there is no evidence that in 1961 people were poorly clothed, or that there was insufficient furniture in their houses, the implication is that the extra time spent shopping is not done out of necessity but out of choice. Now that people are increasingly choosing to spend their time in shops, it is legitimate to consider some shopping at least as a leisure activity.

A broadening of the plane of analysis from the detailed empiricism of Gershuny to the nature of society today makes it possible to identify a line of argument that supports this view of the rôle of shopping. The argument, as outlined by Rojek (1990), draws upon Baudrillard's view and analysis of society. The argument is that mass communication works to make a society whose very purpose is to consume. However, the drive to consume is not to (or not just to) gain the benefit from the object or service but to use consumption to signify something about personal identity. As Rojek (1990: 9) notes "the world of commodity exchange . . . has given way to a more sociologically ambivalent world of the mass circulation of signs". It is the image that the object projects rather than its value in use that matters more. This certainly is the basis on which many cars are marketed and purchased, and certainly, for many, it is a prime consideration when buying clothes. It is the need to identify and acquire these images that shifts some shopping activity into the leisure domain.

Shopping behaviours

If some shopping then constitutes a use of leisure time how can we conceptualize shopping in order to draw a distinction between different kinds of shopping behaviour? The basic distinction lies between shopping as a *functional activity* and shopping as a *leisure activity*. In reality it is better to think of shopping not as two mutually exclusive categories but as a bipolar continuum. At the same time it is not appropriate to see individual shopping trips as having either a functional or a leisure basis. Any single shopping trip can combine a number of sorts of shopping behaviour that reflect a complex of interdependent influences, such as wealth, life style, self-image, motivation and individual valuations of competing leisure opportunities.

The characteristics of functional and leisure shopping are summarized in Table 11.2. Functional shopping is valued because it meets identified needs, while leisure shopping creates wants rather than satisfying needs. The functional shopper has a high expectation of finding what is wanted and ensures this by repeating previously successful forms of behaviour. Functional shoppers value predictability, and their actions are frequently highly predictable, in the choice of shopping centre,

Table 11.2 The characteristics of functional and leisure shopping.

Shopping as a functional activity	Shopping as a leisure activity
High expectation - predictable	Values difference, wants novelty
Meets identified needs	Creates wants
Targeted activity	Browsing, opportunistic
Time efficient	Consumes time

choice of shop and selection of goods. Leisure shopping, by contrast, values difference and the unexpected; it looks for and expects novelty both in products and environments. Functional shopping is structured by routine or planning. The functional shopper targets centres and stores and makes use of established routes between and within stores. Leisure shopping, by contrast, is not route dependent. It is based on browsing, and both routes and purchases are reactions to opportunities and to impulses. The last contrast we can draw between functional and leisure shopping is the way in which shoppers value time. Functional shopping is time-efficient. The object is to make purchases in the shortest time in order to allow more time for optional rather than necessary activities. Leisure shopping, on the other hand, is a means of "consuming" time. Shoppers gain satisfaction and enjoyment from spending their time in this way. It is a pleasurable activity in its own right and it usually takes place in a pleasurable environment.

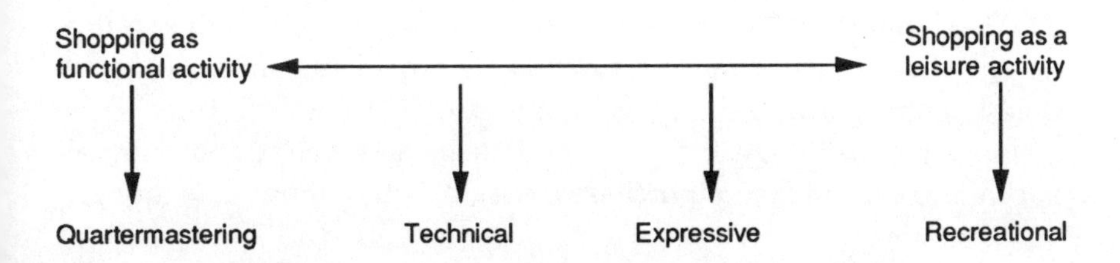

Figure 11.1 Categories of shopping behaviour (after Carr 1990).

Carr's (1990) categorization of shopping activity can be mapped into this functional leisure continuum (Fig. 11.1). She has pinpointed four principal forms of shopping behaviour: quartermastering, technical, expressive and recreational. Her categories of quartermastering and recreational fall towards the extremes of the continuum. *Quartermastering* she defines as the routine purchasing of essentials and everyday items. Its characteristics are that it has to be done, it is boring and "most people see it as a chore" and a "waste of valuable time". A trip to the supermarket clearly falls into this category. *Technical shopping* is the purchase of technical goods that do a job of work: for example, dishwashers, freezers, CD players, video recorders and cars. The purchase decision requires research into the appropriateness of technical operating features, warranties, after-sales service, and price and price variations. Technical shopping shows many of the characteristics of functional shopping as it requires planning, organization and analysis;

the need for the good is established prior to the shopping trip, stores where the shopper expects to find the good are targeted. At the same time it also reflects some features of leisure shopping, particularly the sense of pleasure that comes from purchasing and owning the good. *Expressive shopping,* by contrast, lies further along the continuum towards leisure. It is the purchase of goods through which people can project themselves as they would like to be seen. It is through such purchases that they create an image for themselves. Clothes fall into this category, as do cosmetics and many jewellery purchases. There is also an element of expressive shopping in the purchase of many technical goods. Expressive shopping values such features as design and reputation. These are key signs that provide information about taste and judgement to the rest of the world. It is for this reason that people buy clothes with designer labels, T-shirts to show that they have been on exotic holidays and cars that are "badged-up" from base models. *Recreational shopping,* Carr's last category, is shopping as a pure leisure activity. People decide to go to a shop or a shopping centre as a trip out. Very often the choice of a shopping centre may be influenced by design characteristics: an old town is chosen in the summer, an enclosed centre in the winter, a garden centre on a Sunday. Trips whose main purpose is to remain aware of the range and type of goods on sale in order to inform future purchase decisions fall into this category too.

The circumstances of leisure shopping

The character of leisure shopping is, as discussed above, significantly different from much other shopping behaviour. As Jackson (1991) says, it is important to understand leisure shopping as an experience, to understand more about the structure of the purchase process and the sources of pleasure at each stage. To see the purchasing process merely as an economic exchange is almost certainly to misunderstand it from the consumer viewpoint. Expressive and leisure shopping contains some or all of the following stages: identification, comparison, selection, purchase and then display. Each of these stages can generate pleasure. The anticipation of making a purchase, the evaluation and display of competing qualities, the "celebration" of a purchase with a meal or a drink, and the display of the purchase to others so it can be admired, are all elements of buying behaviour that reinforce the pleasure of owning something.

It is important to appreciate this feature of shopping in order to understand the circumstances in which leisure shopping takes place. There are two components to leisure shopping. The first, that it takes place in leisure time, must always be there; the second, that it is associated with particular environments, is frequently fulfilled but is not strictly necessary. Key times for leisure shopping are weekends, holidays and days out. Purchases are made for a variety of motives: because the opportunity arose, as a memento to prolong the experience of the trip or holiday, as presents to show the recipients where you have been. Sunday shopping in particular is likely to have a strong leisure dimension because in going to shops people have exercised choice about how to fill their leisure time. It is important to know much more about the motives and attitudes of shoppers in leisure time. If these motives and attitudes are different from those on other shopping occasions, every effort must be made to ensure that retailers do nothing to undermine or modify them. A good example of leisure-time shopping is a weekend visit to a garden centre. In this case not only is the time important but it is reinforced by the purchase of leisure-time goods.

Leisure shopping also occurs in particular environmental settings, for example, historic towns. These are visited because of their character. They offer an individual setting for shopping and a greater chance of finding individual retail provision. In this case the shopper is attracted by the environment, which contributes to the enjoyment of the shopping trip. This type of leisure shopping can be described as "ambience" shopping.

The response of the retail system to leisure shopping

The retail system's response to leisure shopping has not been straightforward. It is clear that in some quarters the nature of these developments in shopping behaviour is understood and that this understanding has led to an appropriate retail response. In most cases, though, the response has occurred at the fringe of the retail system. The great majority of developers and retailers have, through their actions, shown either a total failure to appreciate the implications of these new types of shopping behaviour or where they have appreciated that some change has taken place they have often misunderstood it and have responded inappropriately.

Predictability and the issue of consumer frustration

The heart of the problem for much of the retail system has been that it has forgotten that it is a consumer-led business. Certainly the claim will be made that it is responding to consumer needs and wants, but what is ignored is the fact that it is not responding effectively to growing consumer frustration. The evidence for consumer frustration is widespread but anecdotal. For example, letters to local newspapers in St Albans, Hertfordshire, through the 1980s complained of the changing character of the main shopping street, targeting the expansion in the number of financial service outlets, the growth and concentration of fast-food outlets in one section and the number of travel agencies being established. Concern revolved around two matters: reduction in choice and diminution of individuality. These structural changes are described further in a report by the Distributive Trades Economic Development Committee (DTEDC 1988). *The Independent on Sunday* (3 November 1991) argued that the dominance of chain-store retailing was creating a bland shopping environment, the effect of which was to reduce local character as a result of the increasing uniformity of the retail provision. One property professional was quoted as saying that "towns are being transformed so that all they consist of is chain stores. All towns are beginning to look the same."

What is happening in the store cannot be divorced from what is happening in the overall retailing environment. Both have an impact on the shopping experience. From the consumers' point of view the problem is one of predictability and the convergence of stores on the range of goods they stock. As a consequence, shops and shopping centres are beginning to lose their individuality.

The loss of individuality in the retail system is a result of a pattern of change in retailing that has been 60 to 70 years emerging, but has accelerated more recently. These changes are responsible for the growth in consumer frustration and boredom.

The patterns of change are well known:

- the progressive separation in both organizational and spatial terms of convenience shopping from comparison shopping;
- the development of speciality retail provision to reflect and create market segmentation; and
- the expansion of multiple retailing at the expense, particularly, of individual and regional retailers.

Each of these patterns, in its time, was considered "modern" and innovative and thus had a "novelty" value. However, as the innovations became the norm they established an effect not only on shopping behaviour but also on shopping attitudes and shopping cultures. In the case of the development of the supermarket and its extension into superstore retailing, it has driven both a spatial and temporal division between food and household shopping and fashion shopping. In particular, the extension of shopping hours has, for many shoppers, created an expectation that food shopping will take place during the week, especially in the evenings, leaving fashion shopping as the preserve of the weekend leisure period. Superstore retailing has, in this way, played a part in creating a culture of the weekend as a leisure period, and as shoppers we have absorbed the ethos of the superstore and accepted that our purchases of food and household goods should be made in an efficient and functional way.

Fashion chains, have evolved similarly. At first, such stores as Next, Principles, Burton represented innovation and novelty through their targeting of a market sector. They caught the mood of their target markets and allowed people to express themselves through what they wore in a far more effective and imaginative way than before. Yet their success has been their undoing. What was a novelty is, for the next generation, an established part of the retail scene, and what is more it is an established part of the high street nationwide. Their ubiquity has effectively reduced choice and increased predictability.

The overall effect of these changes can be summarized as the process of innovation leading to predictability. In the case of superstore development this appears to have been desirable: the effect has been to highlight the "choice" dimension of the freed-up weekend. However, this choice has proved illusory as innovations in fashion retailing have colonized the shopping centres, increasing the predictability of what will be found and reducing the opportunities for individual self-expression. In 1986 multiple stores made up 70 per cent or more of the retail presence in 11 per cent of the 250 major shopping centres in the UK (DTEDC 1988). The largest proportion was 91 per cent in Brent Cross and East Kilbride. Multiple dominance on this scale produces predictability throughout much of the retail sector and renders shopping a functional, routine activity.

Leisure opportunities in shopping centres

It would be wrong, however, to charge all of the retail system with a failure to appreciate the importance of non-functional shopping behav-

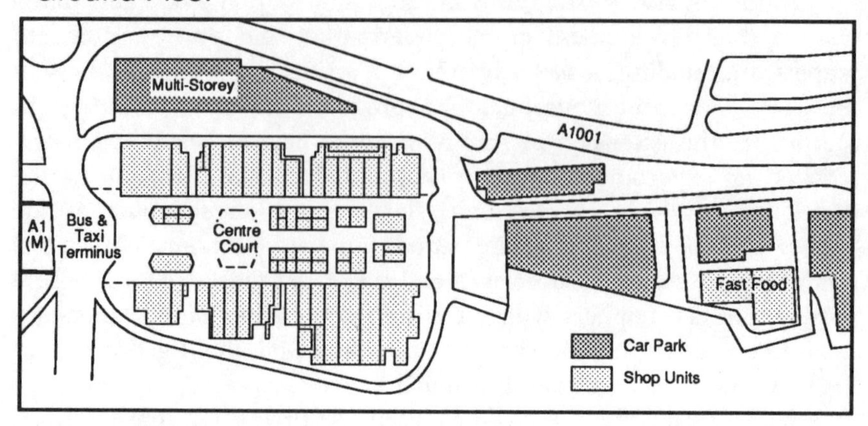

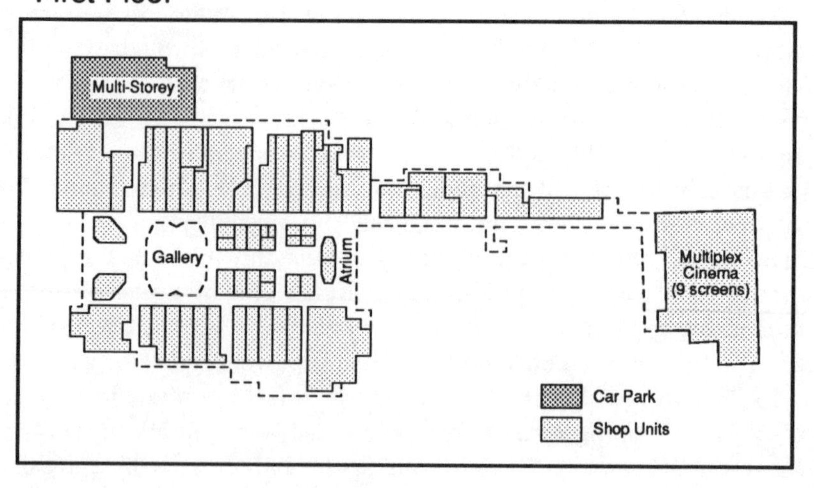

Figure 11.2 The Galleria shopping centre, A1(M), near Hatfield, Hertfordshire.

behaviour. Some retailers and developers have noted the growth of leisure shopping, but by their responses it is clear that they have not totally understood it. Typical of the responses is Liberty 2 shopping centre in Romford. This development is a combined retail and leisure centre built on the edge of the town centre and connected to a 1960s precinct by an underpass. It has Sainsburys as its anchor store, a range of small units around an atrium, and on the top two floors leisure facilities, including a cinema, gaming club, restaurant and night club. A similar concept was tried on a larger scale at the Galleria at Hatfield

(Fig. 11.2). In this case, however, problems with the tenant mix, the layout and design, and the leisure–retail balance (compounded by opening in the depth of recession) led in September 1992 to the receiver being called in.

The theory underlying the development of leisure opportunities in planned shopping centres is that there will be a need to provide formal leisure facilities to meet the demands of an increasingly leisured society. Consumer demand is insufficient to support a 100 per cent retail component in a scheme. Leisure, it is argued, will provide a space user capable of meeting the rents. The theory continues by arguing that the retail component benefits as a result of an enhanced service to shoppers, and in this way the pulling power of the centre is increased.

The result is a *combination centre* with retail provision and leisure provision. Such centres are well documented. In Canada the West Edmonton Mall (Butler 1991, Jackson 1991) has 15 per cent of its space devoted to recreational and leisure facilities, which generate 25 per cent of turnover. In the UK the MetroCentre at Gateshead has been the focus of professional interest and analysis (Harrison 1990, Johnson 1990). The evidence in the UK is that the inclusion of a leisure component at the development stage has increased in recent years, although the scale is still limited. A survey of a sample of centres that started trading between 1988 and 1991 showed that about 33 per cent had a formal leisure component, though none of the centres with under 18,600 m^2 (200,000 sq ft) of retail space had any leisure provision (Newby 1991). The most common leisure provision was multi-screen cinema (80 per cent of those with leisure provision), followed by skating rinks and libraries (33 per cent). Others had theatres, bingo halls, discos and fitness centres. If restaurants and other eating facilities are included as leisure provision then all centres had them. In general, there were one or two restaurants per centre for centres up to 27,900 m^2 (300,000 sq ft), thereafter there was an increase up to a maximum of 18. Forty per cent of the centres had a food court, though they were concentrated in centres above 200,000 sq ft (18,580 m^2). The number of outlets varied between 6 and 10.

Leisure activities and shopping

There is only scant evidence, however, that the co-existence of leisure and retail provision enhances the shopping experience. For one thing, shops in most of the centres were closed when leisure facilities had their greatest draw, so the opportunities for synergy are limited. Kaye (1988) points out that the average shopper would be unlikely to repeat the experience of combining leisure and shopping after leaving frozen food in the car for three hours while visiting the cinema and then having to clean the car out. Second, the leisure opportunities tend to have a long duration and are exclusive of other activities. This reality has tended to be recognized by developers and so, for example, in Meadowhall, Crystal Peaks (Sheffield), Lakeside (Thurrock), Liberty 2 (Romford) leisure and retail are spatially separated. In most combination centres the layout and appearance of the retail component owes more to the design concepts of the 1970s than to those that might reflect shopping as a leisure activity. In Lakeside, for example, only in the festival market in the leisure pavilion is there any attempt to create individuality.

Food courts

It is with their food courts that combination leisure–retail centres come closest to enhancing the shopping experience and creating a leisure dimension. Food courts provide an opportunity to rest and take a break. In this way they reduce the stress of shopping, although the furniture is usually designed so that after 30 minutes, or an hour at the most, the stress of shopping is usually preferable to the discomfort of sitting. But the return to shopping is usually a return to functional as opposed to leisure shopping.

Environmental enhancement

The need to move beyond the combination of leisure and shopping facilities in one location has been recognized by Lichfield (1990). Her view is that shopping behaviour is moved into the leisure domain when stress is removed. She offers solutions that will provide rest, relief and distraction. The retail system has moved in this direction by introducing entertainment into the shopping environment. However, while the leisure component is a developer's initiative, the enhancement of the shopping experience through entertainment is the result of centre-management initiative. A survey of centres opened between 1988 and 1991 showed that all centres offer some sort of entertainment (Newby 1991). The most popular activities were live music (89 per cent of centres), variety acts (72 per cent of centres). Most other activities (dance, street theatre, concerts, buskers) were found in between 33 per cent and 44 per cent of centres. There was, however, great variety in the frequency with which entertainments were provided. Live group music performances occurred from as rarely as only at Christmas to as frequently as every Saturday. Variety acts occurred with more frequency. The type of act engaged creates a sense of "circus in the mall" – clowns and jugglers were the most frequently mentioned plus Punch and Judy shows. Centres over 18,600 m^2 (200,000 sq ft) offered a wider range of activities, including concerts (particularly at Christmas), folk-dancing, jazz festivals and fashion shows. As well as offering entertainment, 40 per cent of the centres tried to create a "theme shopping" experience when traders dress in a particular style – St Valentine's Day, Easter, Halloween and Christmas were the most popular periods for these (Newby 1991).

The provision of entertainment to this degree by modern shopping centres indicates that, recession apart, centre managers recognize the need to offer inducements for shoppers that enhance the quality of the shopping experience. The evidence, however, is that entertainment is seen as an adjunct to the shopping trip rather than an integral part of it. For entertainment to be successful requires a deeper analysis of its rôle and purpose in the shopping trip.

The tourist dimension

It would be wrong, however, to portray the response of the retail sector to the wish for leisure shopping as wholly one of missed opportunities and misunderstandings. There are situations in which retail provision meets the desire to use shopping as a leisure experience. Some of these have already been touched on, such as, for example, the enhancement or colonization of a heritage environment by retailing in response to the opportunities offered by a tourist market. Jansen (1989) explores the elements that contribute to central Amsterdam's being a "fun shopping" environment. "Atmosphere", he found, was all important and the things that contributed most to it were street musicians, artists, crowds, the market and goods of unusual provenance. These elements are also found in the UK. Shoppers enjoy the ambience, texture and grain of such shopping centres as Canterbury, Norwich and Chester. Retailing can even play a part in reclaiming such environments, for example, developments in Tunbridge Wells and South Street Seaport, New York.

However, the pressures towards reduction of individuality are present here too and, in tourist environments in particular, the growth of gift shops stocking the same range of goods has the potential to shift shopping back along the continuum from being a leisure activity. Some of the most successful retail "colonizations" of heritage environments have been made by antique shops. Table 11.3 shows the retail "hierarchy" for the antique sector. Clearly, there is a relationship with the size of the market, but equally there is an very strong link with urban character.

Table 11.3 The hierarchy of antique centres.

Number of antique stores	Number of centres	Centres
Over 1000	1	London
100–999	1	Bath
50–99	7	Bristol, Woburn, Newton Abbot, Stow on the Wold, Brighton, Birmingham
30–49	7	Bournemouth, Cheltenham, Tunbridge Wells, Norwich, Newark, St Leonards Harrogate
20–29	20	–
15–19	21	–

Source: Calculated from Adams 1992.

Antique shops offer goods dramatically different from those offered by chain stores: varied ranges of stock and items each of which has a high degree of individuality. Because shoppers do not know for certain what will be stocked, each shopping trip is low on predictability. Exploration of the stock consumes time (which is one of the objectives of leisure shopping) and there is always the excitement of finding a bargain, a rare item or something that has personal or sentimental value. The purchase process also possesses features that support leisure aspects of shopping. Discussions about the provenance of an object, particular design features, its age and quality all use up time, allow the buyer both to learn and to show off knowledge and, perhaps most importantly, create or add "value" to the object. This whole process is important in reaching a decision to buy. Here again negotiations about the purchase price are different from the way in which we buy in the rest of the formal retail system and there may be occasions in the decision to buy in which the discount or "saving" is as important as the purchase price.

Festival shopping

While shoppers' preferences for heritage environments create the preconditions for one type of leisure shopping, it is clear that the retail sector itself has responded in a particular way to the wish for leisure-shopping opportunities. This response is typified by what has become known as festival shopping. The elements of this have been summarized by Ogg (1990: 2) as "greenery and activity; colour and delight; striking retail presentations, small scale selling; people promenading; people relaxing; bands playing; people entertaining; people relaxing; people eating". Clearly once the physical ingredients are in place and successful, the social rôle of the shopping centre becomes an important source of attraction in its own right. Festival shopping is increasingly found as a small component of planned shopping centres but it has been particularly successful where it constitutes an overall theme for a shopping environment, such as the Piece Hall, Halifax, or Covent Garden, London. In the latter the range of shops, the mix of places to eat, the opportunities to see street theatre together with a unique environment combine to make this an archetype leisure shopping complex. Again, though, it contains the seeds of its own destruction as rentals move up to levels that can only be met by multiples. At an earlier stage of development than Covent Garden and with the same potential for success is Camden Market in northwest London. Here again the elements for leisure shopping exist: a canal-side location whose warehouses provide a distinctive environment, individual retailing (often with a craft basis that produces variety in products sold), and a large number of food stalls and entertainment. The weekend market has now expanded beyond its original confines into adjoining derelict industrial property, south into Camden Town and north towards Chalk Farm. Its commercial success is reflected in the investment that is taking place and has taken place in refurbishing premises.

Informal markets

There is, however, a stage of retail provision back from Covent Garden and Camden Town that is built almost entirely on leisure shopping and has grown appreciably in recent years: the weekend market and the car-boot sale.

Neither of these are usually found in quality environments but what they lack in this respect they make up for in other ways. They are placed squarely in leisure time. They offer variety and uncertainty. Most people do not visit them out of need. Data collected from the *Herts Advertiser* on the numbers of car-boot sales advertised in St Albans reveal the considerable increase in these events. For two months of the year in 1983 there were 10 car boot sales. By 1987 there were 21, and in 1991, 25. The increase from the early 1980s indicates that car-boot sales are now an established, if small, part of the retail scene that caters for the demand for leisure shopping.

The prospects for leisure shopping

Not every shopping environment can or should be converted to accommodate leisure shopping. Much of our shopping behaviour will continue to be functional. The leisure element in buying new clothes for children is always likely to be lower than when buying oneself a new outfit. However, leisure shopping is important because it lies at the margin with other forms of leisure. A shift one way could move a profitable retail segment into other parts of the leisure market. On the other hand to understand leisure shopping and to respond to the opportunity with appropriate retail provision could draw people from other forms of passive and active leisure and create a more secure basis for a currently hard-pressed retail sector.

The areas on which action should be focused are the link between leisure and retail trading, the strategy for entertainment, the exploitation of leisure time and the rôle of design and design management. It is not, however, just developers and shopping-centre managers who should explore and exploit the opportunities here. It is equally important for the health of the high street and other traditional shopping environments that the place of leisure shopping in their plans is considered by town planners, town-centre managers and the multiples. The Distributive Trades Economic Development Committee report on the future of the High Street (DTEDC 1988: 7) identifies the need to develop retailing "which gives distinction to the shopping area" and "by incorporating such uses as leisure and/or entertainment services". More than anything, traditional shopping centres require collaboration and co-ordination between councils as managers and retailers as providers if their economic health is to be sustained.

There are four areas for developers and centre managers to focus on: the link between leisure and retail, the strategy for entertainment, the exploitation of "leisure-time" shopping, and the place of design.

The combination of leisure and retail

The combination of leisure and retail is possible in both established and new shopping centres. In the former it is more likely to be based on the provision of food and drink and musical entertainment. In new shopping centres the opportunities are greater. However, in both situations it is difficult to combine both functions into a supportive system. If leisure and retail draw their custom at the same time there is potential conflict. Most developers have therefore used leisure as a means of prolonging the shopping trip and extending the trading day. The challenge for developers is to find the combination that yields most profit. It may be that the balance has to shift so that leisure becomes the principal draw and the centre acquires its distinctiveness on this basis. This certainly works on a small scale, for example the National Trust shop, located at the exit of the property, which is used to prolong and enhance the leisure experience. The weighting towards leisure may also be successful at the larger scale. The Mall of America (Bloomington, Minneapolis) will provide a test of this new leisure/retail concept. The leisure draw is more co-ordinated than that found in West Edmonton Mall. Its core is a 3 ha (7.5 acre) air-conditioned theme park with a range of rides and entertainments organized and marketed around the theme of Snoopy. The theme park is supported by an 18-screen cinema, ice rink, mini-golf course, health club and hotel. Around this is 241,500 m^2 (2.6 million sq ft) of retail space. The market to support this complex has been estimated at 27.4 million people living within 400 miles. Leisure provision on this scale raises the possibility of shopping visits that are measured in days rather than hours.

Developing an entertainment strategy

The development of an entertainment strategy is perhaps the area with the greatest potential. At present there is provision but no real sense of the rôle it should play. It exists as a marginal activity, not in its own right with a clear purpose in enhancing the shopping experience. Much of the entertainment provision revealed by the survey of current practice was ad hoc. Good practice requires centre managers to do two things. First, they need to plan a programme over a trading year; then to monitor and evaluate it to ensure that it is achieving its objectives. Secondly, they need to identify what they want entertainment to achieve; whether, for example, it is to act as a crowd-puller or is designed to reduce the stress of shopping. How long do they want shoppers to be entertained? Does the centre want to become known as a place of street entertainment? Is the entertainment to be provided free, as is usually the case now, or can the entertainers ask for money? The answers to these questions are necessary in order to determine the right space allocation, the most appropriate location and to decide whether seating is required. Ideally these questions need to be asked at the design stage so that the right solutions can be built in. If entertainment is to play a significant rôle then an auditorium facility must be provided. The problem managers face is that there is no research on which to base an answer to these on a "fill the space and fill the time" basis whereas what is required is for centre managers to approach it as a professional responsibility. There is almost certainly a need for training in the development and implementation of an entertainment strategy. If they are to respond effectively centre managers not only have to know the market as a whole but also market segments and they have to understand what appeals in entertainment terms. To their other professional responsibilities the centre managers who aim to develop leisure shopping must add the rôle of impresario.

The exploitation of leisure time

The probability that Sunday trading laws will be revised provides a real opportunity for the retail sector to capitalize on leisure time. If leisure time (Sundays, late evening) is colonized for shopping then retailers would be unwise to provide more of the same. These periods need to be distinctive. If leisure-time shopping means just the extension of retailing hours, then it will ultimately be a wasted opportunity. At first shoppers may enjoy the excitement of visiting stores at unusual times. Retailers, however, may find that it does little more than spread the shopping "spend" over more trading hours. Indeed should boredom with the range of goods and mix of shops increase, the situation might even deteriorate. Retailers will then look back and give thanks that a combination of English legislation and pressure groups such as Keep Sunday Special prevented them from wasting a valuable source – consumers' leisure time. Retailers and centre managers need to consider how they can create a distinctive character for leisure-time shopping. Certainly an entertainment strategy is a strong card to play. They may respond by creating a carnival atmosphere including, for example, a sequence of acts or street theatre, mall trading, or a wider range of refreshment and recuperation facilities. The problem, however, facing many centres is restriction on activities that might create fire hazards. This issue needs to be considered at the design stage. The rôle that entertainment has to play is to attract people from other leisure activities. For this reason entertainment has to be the draw, not the shops. The entertainment needs to be of a volume and quality that guarantees that going to the shopping centre is a real alternative to going on a visit, to a sports event or even watching television. Retail has to trade off the leisure experience; the shopping opportunities perform the same rôle as the National Trust shop, restaurant and tea room after the country-house experience. It may be that in order to generate leisure shopping in leisure time shops will need to accept that it is marginal to their trading week and generates a lower per capita spend.

Environment and design

Modern shopping-centre design has played a significant part in making shopping a functional activity. Layout and surface design emphasize efficiency. Shiny surfaces lack depth and texture and reinforce the immediacy and thus the superficiality of the experience. Leisure-shopping environments frequently lack order, or at least hide their order. Nowhere is this clearer than in markets or heritage-shopping environments. In the former the concentration of goods and a degree of disorder in their display creates a sense that there may be bargains to be had. Shopping then is not a functional activity but an exploration. In older towns shoppers have many options in terms of routes and they can feel enticed to go one way rather than another; this contrasts with the conventional shopping mall.

Leisure-shopping environments are often experienced as a sequence of spaces in which the environment may be as much of an attraction as the goods in the shops. Centre design should recognize this and must be a major area of research and development if retailers are to create a leisure-shopping experience. The task for designers is to create variety and mystery rather than immediacy; they must produce enclosure as well as openness; they must create an environment that encourages people to explore rather than one in which they see their goal in the distance. Perhaps the models they might use lie in the past rather than in the 21st century? As an example of this we might point to Brewer's Quay in Weymouth where, in refurbished industrial buildings a medieval streetscape has been created that mixes a trip through 600 years of Weymouth's history with 1,860 m^2 (20,000 sq ft) of specialist retailing.

Conclusions

Retail developments, it is argued, are helping to turn the bulk of mainstream shopping into a functional activity. Retailers understand the importance of novelty, but their systems, even down to stock control, which cuts out all but the fastest-moving lines, seem more intent on delivering "sameness": the same goods, in the same shops, in the same high streets and shopping centres. Eventually consumer boredom and frustration will set in. Some retailers and developers understand the challenge: to develop shopping as a leisure activity. For example, Sutcliffe (1988: 63) looks back to times when "shopping used to be fun", pointing to the market as a social occasion and orchestral performances in shopping arcades in Victorian times. The "gut feeling" of the property professional has not been matched by the intellectual interest of academics. Academic analyses are rooted in identifying the bases on which rational decisions are made. Because of this there has been a failure to understand that "fun", too is a commodity. It is an area of omission that should be addressed. However, in terms of an immediate practical response, the combination leisure–shopping centre probably has less potential to develop leisure shopping than do centre managers with their opportunities to enhance the shopping experience. The entertainment-based response, however, frequently lacks a strategy and an understanding of how it can be used to prolong the shopping trip and increase expenditure. The likely collapse of Sunday trading legislation represents a real opportunity for centre managers to create a leisure shopping environment. In the long run, though, the solution lies with developers, who must respond with alternative and idiosyncratic design solutions which re-establish uniqueness as a characteristic of the shopping environment.

CHAPTER TWELVE

The Greening of shopping

Frank W. Harris & Larry G. O'Brien

Green consumerism acquired an outstanding political platform in the Global Forum that paralleled the 1992 United Nations Earth Summit in Rio de Janeiro. An army of environmentalists, aid organizations and non-governmental development agencies met to discuss an environmental agenda much broader than that of the official conference. One of the important topics was Green consumerism: the development of Green awareness in day-to-day shopping and life-style decisions. Its fundamental premise is that the health of the planet rests more with the individual than with governments or multinational companies. Individuals lie at the sharp end of marketing channels that link consumers with producers and, as such, would represent a considerable commercial power over the whole channel were they only to act in consistent, environment-friendly ways. The forum and the international media coverage it received enabled Green consumerism to publicize its most central point which, put simply, is that if individual consumers commit themselves to consuming in environment-friendly ways, the international capitalist market for goods and services will be forced to provide them. In this way capitalism becomes Green capitalism.

For some Greens the processes associated with capitalism are seen as the causes of environmental decay and therefore in their opinion no amount of technological tinkering with vulnerable ecosystems and complex, dynamic systems such as the atmosphere can be tolerated. Other Green consumers combine their concern for environmental health with a concern to protect the standards of living experienced in the West. To feed the population, raise standards of living and increase the quality of life, many Green consumers accept the need for technological solutions so long as they are environment-friendly. There is thus an inherent heterogeneity among Greens and scope for contradictions and exploitation in both consumer and retailer–producer responses to Green issues.

This chapter considers the spectacular rise in Green awareness during the 1980s and its implications for the behaviour of consumers and retailers. Within the British context, it explores what evidence exists to indicate that Green consumerism is bringing about environment-friendly attitudes on the part of retailers and producers. The Greening of shopping is seen as an issue of widespread concern and contradiction. However, the trend developed in a period of economic optimism during the 1980s and this has since been replaced by recession, financial instability and a "Green fatigue" among many sectors of the public. In the early 1990s the future of Green retailing is less certain, but there is still considerable public interest and a growing body of environmental legislation.

Perspectives on environmental health

Public concern for the environment is not new and has affected many societies. Of particular importance is the nature of the relationships between human populations and the physical resources needed to support their needs and ensure their survival. The specification of a formal, law-like relationship between people and the resource base was outlined in the so-called "scientific" studies of Thomas Malthus in the 18th century. Crudely reduced to two laws of nature, the Malthusian belief that population growth inevitably outstrips the capacity of the land to support it, envisages a world of "natural" human disasters, with populations being kept in check by disease, pestilence and war. Leaving aside the many technical and procedural objections to this position, it remains true to say that Malthus's work continues to be highly potent in contemporary debate, particularly in the work of Ehrlich and those proponents of a philosophy of sustainable development (Ehrlich et al. 1973, Meadows et al. 1972).

In contrast with this is the belief in inevitable technological advance; a view that seeks its justification in the assumed progress of the industrialized West, which has delivered life styles marked by standards of comfort, convenience and conviviality that are now taken for granted. Economists such as Beckerman (1992), argue that the pessimism built into the Malthus model simply arises from the failure of commentators to acknowledge humanity's ingenuity in replenishing or otherwise reevaluating the environmental resource base. It is therefore feasible to assume that the life styles typical of the West can be sustained into the future, perhaps for ever.

For such technocentrists, the mechanism that capitalizes most effectively on this assumed ingenuity is the existence of free economic markets; that is, markets that are not directed by any single vested interest and that achieve their efficiency by a price mechanism linked inextricably to a philosophy of enlightened self-interest. Economic actors such as companies, managers and workforces are assumed to behave so that their best interests are maintained. If everyone does this, everyone benefits and the resources needed to produce the goods and services needed for the process are assumed to be used in the best possible way.

To Ehrlich and others of his persuasion, this is arrant nonsense. First, the evidence of industrial capitalism is the result of such a short time span that it hardly deserves the label of a "natural" process. Secondly, the technocentric approach fails to recognize the special circumstances of industrial capitalism which made its gains for the West on the back of large external empires, cheap labour forces in those empires, and imperialist policies that transferred wealth from one part of the world – the so-called "South" – to the industrial economies in Europe and the USA. The concept of unequal resources remains today in spite of notable shifts of commercial power to Japan, South Korea, Singapore and other "Southern" nations (Taylor 1991). Thirdly, even if economic indicators do suggest that living standards and the quality of life are improving across the planet as capitalist processes become more globalized, these can hardly be considered objective measures about which there is consensus among commentators. They merely reflect the underlying assumptions and accountancy procedures of capitalist economics. The hard "objective" measure of the contemporary Malthusians is the ultimate failure of the planet to support its population, irrespective of humanity's ingenuity in changing the ground rules.

These two perspectives represent the opposite ends of what might be termed the Western approaches to environmental health. They are Western in the sense that they arise from perceived Western economic and demographic experiences and represent Western views on biology and economics. They are not the only views being aired, but they seem to represent the key tensions in the current debate. For many people concerned with the future of the planet, the pull between biological controls and economic ingenuity represents the core of the problem. It is in this context that a growing Green consumer awareness needs to be considered.

Definitions of "Green"

"Green" has been used in a plethora of ways (Porritt & Winner 1988, Dobson 1990, Yearley 1991). Although most frequently used as a term associated with a growing awareness of the environmental consequences of human activity, "it seems clear that whereas a concern for the environment is an essential part of being Green, it is . . . by no means the same thing as being Green" (Porritt 1986: 5). The expression should therefore more accurately be seen as an umbrella term covering many philosophical and ideological areas including social and moral responsibility, ethics, aesthetics and considerations of equity. Not only is this lack of a clear focus problematic for Green activists and politicians, who are sometimes perceived as being "a ramshackle coalition" (Porritt 1987: 24), but it also offers the potential for extremely loose interpretation across a whole spectrum of human activities and behaviour. As Porritt & Winner (1988: 9) note, "Green" has been used to refer to: "a rich and extraordinarily diverse movement with new shoots springing up all over the place. Activities as diverse as becoming a vegetarian, signing a petition to stop a motorway or living under a plastic sheet outside a cruise missile base in Berkshire are all held to be 'green'. Among a plethora of different aims, the most commonplace is simply to improve the quality of life: to be healthier, to save a lovely old building or to protect the countryside". Yet irrespective of this variety of definitions, Green consumer movements began to emerge during the 1980s in many Western countries. Their aim is to use the power of competitive free markets to redirect commercial activities from environmentally destructive forms of production and consumption into forms that sustain the planetary resource base while simultaneously maintaining the comfort of Western life styles.

The emergence of Green retailing in the UK

Every month since the early 1980s, the market research organization MORI has been asking the British public two unprompted questions concerning its opinion of the major issues facing contemporary Britain. For years the traditional responses included comments such as law and order, national security and the state of the economy. However, from about 1988 a significant proportion of respondents began listing the state of the environment as the most important issue. On three separate occasions more than 30 per cent of the respondents listed the environment as their first or second priority, with the peak of 35 per cent being reached in July 1989. No doubt much of this was fuelled by media coverage of issues such as ozone depletion, global warming, acid rain and the destruction of the rainforest. Events such as the 1986 radiation leak from Chernobyl and the 1989 oil spillage from the *Exxon Valdez* in Alaska all served to heighten the level of environmental concern, especially among children. These and other disasters, combined with the publicity surrounding such agreements as the Montreal Protocol on CFCs, or United Nations reports such as the 1987 Brundtland report, have raised general public interest in environmental matters and fostered a concern for public participation in saving the planet.

The roots of the British Green consumer movement can be traced back a long way (O'Riordan 1984), but for many consumers it was during the 1960s that increased consciousness led on to political activity. One of the key areas of awareness was that capitalist production processes are frequently environmentally damaging, especially where they lead to a reduction in biodiversity through monoculture, forest clearance and the profligate use of pesticides and fertilizers. Furthermore, the negative externalities associated with capitalism, for example, water and air pollution, land despoliation, the creation of waste and the increased use of non-renewable energy resources, were also recognized and challenged, especially as the costs associated were rarely included in the costs of producing goods and services for economic markets (Mishan 1990, Anderson 1991). Organizations such as Greenpeace and Friends of the Earth were founded to focus political dissent against environmental degradation. However, although effective in many ways, such as in campaigns to save the whale or to dissuade consumers from buying clothing made from skins of endangered species, they were unable to force the environment to the centre of the political stage.

However, a turning point was reached when Prime Minister Margaret Thatcher astutely judged the mood of the times by declaring in her speech at the 1988 Conservative Party annual conference that, "we Conservatives are not merely friends of the earth – we are its guardians and trustees for generations to come". Although opponents accused her of attempting to make political capital from what they saw as her cynical hijacking of the fashionable Green issue, her pronouncements served notice that the environment was well and truly on the political agenda. If further evidence were required, it was provided by the 1989 elections to the European Parliament when the Green Party gained 15 per cent of the votes in the UK. Although the electoral fortunes of the

Green Party have dwindled dramatically since 1989, culminating in the embarrassing disarray of the 1992 annual conference, Green issues remain significant in the UK.

In the 1990s there is a much increased awareness of the environmental spillovers and deleterious effects on the biosphere generated by technical developments and human activities most obviously associated with the patterns of production and consumption of advanced industrial countries. As the relationship between producers and consumers plays a pivotal rôle in the economic system, it is self-evident that the retailing industry is fundamental in orchestrating the relationships and transactions involved in the marketing channel. Consequently, it is hardly surprising that Green issues should have become prominent on the high street. However, the wide range of possible uses and broad connotations of the term "Green" has led to considerable freedom of interpretation and application of the Green label. The extent of such heterogeneity has allowed producers, retailers and consumers considerable licence to declare their behaviour as being Green and to capitalize on this for their own ends with little fear of contradiction.

Industry responses to environmental concern

Although the causes are far from simple, "the most important driving force behind our environmental crisis is the sheer size of the throughput of energy and materials through the economy" (Irvine 1989: 90). Implicit in this straightforward statement are the interrelated elements of economic growth, resource depletion and a deteriorating environment that is coming under increasing strain by being thoughtlessly used as a waste sink (Mishan 1990). Central to this interrelationship is the logic inherent in the dynamics of capitalism (Harvey 1982). Investment in a capitalist economy is driven by the pursuit of profit with the intention of accumulating capital through the appropriation of surplus value from the production process. Sayer (1985) argues that capital accumulation is actually a condition of survival of the firm. Commercial pressure requires producers constantly to increase output and for retailers to sell more each year.

To ensure continued success in the marketplace, firms need to have a steady flow of products at various stages of development, as is implied in the product life cycle model. In short, "the essence of the product life cycle is that the growth of sales of a product follows a systematic path from initial innovation through a series of stages: early development, growth, maturity and obsolescence" (Dicken 1986: 98). If the launch of a new product is successful, then its output will be rapidly expanded in order to exploit the growing demand. Eventually, the product will reach maturity as demand levels off and ultimately sales will tail off as it reaches obsolescence, to be superseded by something more innovative. Although all our purchases will eventually wear out, it is also a fact that producers' and retailers' interests require them to sell goods that become obsolete more quickly than they wear out in order to be assured of a future market. Nowhere is this tendency better exemplified than in the fashion-clothing sector where, twice a year, design features such as colour combinations, hem lengths, lapel widths and fabric types are routinely changed, overtly to take account of innovations in style and creativity, but covertly in order to maintain and enhance the prospects for the industry.

Clearly then, the cult of economic growth is a deeply embedded ethos in the capitalist system. Indeed, "to economists, journalists, historians, politicians and the general public, the rate of economic growth is a summary measure of all favourable developments in the economy . . . The presence or absence of growth is looked upon as an indicator of the success or failure of economic policy" (Usher 1980: 1). Not surprisingly, Gross National Product is still the most frequently used measure of the health and buoyancy of national economies although the obvious deficiencies of such an approach are well recognized (Ekins 1986, Porritt 1990). Although it cannot be denied that economic growth materially improves the lot of many, until recently inadequate attention has been paid to the environmental costs of growth or to the efficiency and distributional equity of the system.

Critiques of the economic growth process can equally well be levelled at communism as at capitalism because both systems are "dedicated to industrial growth, to the expansion of the means of production, to a materialistic ethic as the best means of meeting people's needs and to unimpeded technological development" (Porritt 1986: 44). Indeed, the centrally planned economies of eastern Europe were probably guilty of greater environmental degradation than Western market economies (Peet 1991). However, the replacement of state planning by market capitalism, free enterprise and the unleashed materialism of an embryonic affluent society leads to the conclusion that the drive for growth will merely be reorientated.

It is against this backcloth of a growth culture in which the prime aim is the pursuit of profit that producers and retailers have been responding to the challenge of a growing Green awareness. Because of the very vague and imprecise definition of what exactly constitutes Green behaviour, firms in the retailing industry have had a significant amount of freedom to interpret it as they see fit. They have been selective about identifying items from the Green agenda that suit their business development planning. These can then be packaged and marketed in such a way that a convincing Green image is presented to customers (Peattie 1992).

Firms are naturally eager to capitalize on this newly acquired virtue and so have endeavoured to establish their Green credentials firmly in consumers' minds. By resorting to various strategies to achieve this end, they have also provided the researcher with a range of material that can be used to analyze the Greening of retailers. Advertising material increasingly incorporates Green imagery, while Green claims for products are proliferating. Information pamphlets detailing retailers' policies on environmental matters are becoming more widely available. In their enthusiasm to establish their credibility, retailers readily issue press releases to inform as wide an audience as possible of the latest addition to their environmental policies and range of Green products. From all these sources of information it is possible to identify five principal motives lying behind retailers' forays into Greenery.

Profitable market niches

For many retailers, profits can undoubtedly be made by espousing the right combination of Green elements. Independent retailers of organically produced vegetables and wholefoods fit neatly into this category. At the other end of the organizational scale the major supermarket chains have been quick to adopt many Green initiatives. Tesco is amongst the frontrunners, with policies that include expanding its list of environmentally benign products; rationalizing distribution networks; reducing energy consumption; recycling waste paper; environmental improvements and reduction of pollution by, for example, switching from using CFC gases (which damage the ozone layer) in refrigerators and freezers. The Body Shop has been hailed as one of the most environmentally sensitive businesses, with an impressive array of Green policies in place. For example, products are sourced where possible in a non-exploitative way from the Third World. The company is against animal-testing and extensive use is made of biodegradable natural ingredients. The Body Shop does not advertise its products, uses minimal packaging and offers a refill service. Additionally its shops are used as an arena of education in environmental issues.

Marginal policy shifts

Some firms only make a slight change in business behaviour before judging themselves qualified to label themselves Green. Fuel companies make great play of the fact that they have added unleaded petrol to the range of products available on service station forecourts. For some firms the addition of a range of natural cosmetics or biodegradable detergents is put forward as a demonstration of a commitment to environmental issues. Similarly, the confectionery manufacturer Thorntons, which sells loose chocolates and toffee from its shops, is reported to be switching to recycled paper bags and introducing a range of confectionery free of artificial colours and flavourings (*Green Magazine* 1991).

Cost-saving strategies

Businesses are constantly seeking ways to cut costs. Today, although this may be the main aim, an additional bonus is gained from the fact that a firm can also promote such economies as evidence that it is working to reduce environmental impact. It is increasingly common for retailers to carry out internal audits to see where savings might be made. Outcomes may include lower-energy lighting, more efficient insulation, recycling of energy by using heat pumps and more fuel-efficient vehicles. Sainsbury (1992), for example, claimed that its 1992 supermarkets typically used only 60 per cent of the energy that similar stores would have used in 1982.

Cynical exploitation of Green awareness

Many examples can be cited of manufacturers and retailers that jumped aboard the Green bandwagon in a particularly cynical way. Recognizing the potential profitability of appearing to be environmentally aware, these companies make the barest minimum of adjustments to products and policies in order to capture market share. There is much evidence that some firms have made misleading or dubious claims about their products, and fears have been expressed that, "the Green sell could easily degenerate into a 'green con'" (Yearley 1991: 99). Very often, all that has been undertaken is a subtle relabelling exercise. At the simplest level, it has been shown that "products in green packets sell significantly better than similar products packaged in any other colour. In this context, green has rapidly become the colour of capitalist energy and enterprise" (Dobson 1990: 140). Moreover, words such as "green", "wholesome", "natural", "friendly", "traditional", "harmless" and "safe" appear with increasing frequency on product labels.

Among other uses, ozone-depleting CFCs were used as aerosol propellants for products such as perfume sprays and deodorants. However, certain companies that have switched to pump-action vaporization sprays and roll-on applicators now feel it necessary to advertise the newly packaged product as being ozone-friendly. Although items such as toilet paper and manila envelopes had contained recycled paper for some time, as soon as recycling came into vogue, astute relabelling took place, solely as a marketing ploy, in an attempt to persuade purchasers that they had bought a substantially different product whereas it was identical to what they had bought before.

Public relations exercise

Simply put, in the current climate of public opinion, it is good to be seen to be Green. For each of the motives listed above that are driving the Greening of retailers there is also usually an associated publicity exercise of some description, designed to make sure that the customer is fully aware of the retail industry's commitment to saving the planet (Porritt 1990). Whether in the form of cleverly contrived advertisements, information leaflets or even education packs, most retailers will try to make some promotional capital out of their various Green strategies.

Although retailers' responses to the growing importance of Green issues may be classified under the five broad headings above, it should be noted that they are not mutually exclusive and that firms' behaviour may include various combinations. Thus, a firm operating in a Green market niche could well also be involved in a cost-pruning exercise and might wish to gain some promotional benefit from its various environment-related activities.

The question still remains as to whether, under capitalism, retailers can ever really achieve a meaningful reconciliation between profit and a substantial commitment to Green issues. It is unfortunate that the phrase "environment-friendly" has become so commonly used by the retailing industry. Quite clearly, it is virtually impossible to consider any form of productive activity as being entirely environment-friendly. The production of goods for sale in shops obviously necessitates extraction, processing, transport and distribution - all of which consumes resources and energy and generates waste. Surprisingly though, "The Green Consumer Guide" uses the phrase extensively (Elkington & Hailes 1988).

Even retailers that are considered to be Greener than most merit scrutiny. So much has been written about The Body Shop that it is possible to become intoxicated by the image of perfection (Roddick 1991, McKay & Corke 1986). For a more balanced view, attention needs to be paid, for example, to the fact that the company is a multinational organization with all that this implies for energy and resource use. Furthermore, the original range of 15 products has now been expanded to over 250, precisely the sort of behaviour that the above-mentioned product life cycle would have led one to expect of a fast-growing firm. As Dobson (1990: 142) rather pithily observes: "The Body Shop strategy is a hymn to consumption . . . they urge people to wield their purchasing power responsibly rather than to wield it less often".

Serious attempts on the part of retailers to mitigate the environmental impact of their activities are certainly to be welcomed. Though recognizing this point, "deep Greens" remain highly critical of Western consumer society that is inextricably linked to the treadmill of economic growth and driven by an insatiable desire for possessions and ephemeral fashions.

Consumer responses to environmental concern

Economic growth is considered to be vital to the functioning of the capitalist economic system, but it has been beneficial also for the consumer (Bosquet 1977). The conjunction of economic growth, technological advances and increases in the scale of production has undoubtedly enriched consumers quite dramatically in a materialistic sense. Perhaps this rather begs the question as to whether there is therefore a tacit acceptance by many consumers of the inevitability of the environmental consequences resulting from the large scale of production and types of processes required to deliver the goods at the right price. In any event, rising affluence has led to a transformation in the standard of living for many and this has been accompanied by ever-advancing material expectations and growth in consumption (Ponting 1991). In addition, one cannot escape the fact that people gain tremendous pleasure and satisfaction from the purchase and consumption of goods and services. So how realistic is it to assume that the consuming public, increasingly well schooled in the practice of conspicuous consumption, will develop a significant rôle in reducing the environmental impact of production and consumption?

Two of the chief protagonists of the Green consumer movement argue that, "in today's consumer-orientated society the high-spending Green consumer can pack more environmental punch than almost anyone else" (Elkington & Hailes 1988: 4). This notion of power over what is produced and sold being determined by the choices made by sovereign consumers in free markets is fundamental to their argument. Indeed, "the central thrust of Green consumerism is the rejection at the shopping counter of 'flawed' goods and services in favour of more benign ones" (Irvine 1989: 88). Alluding to this alleged power, the Body Shop states that it "meets the real needs of real people" (Body Shop 1991: 1). However, such views should be regarded with considerable scepticism because, if indeed a free market ever existed, today it is a myth rather than a reality: "it was taken over long ago by big business, giant technology and the advertising industry as the determinant of what people shall be allowed to choose from and what they shall be taught to think they want" (Thomas 1978, cited in Dicken & Lloyd 1981: 81). With such powerful external influences over the choices they make, consumers are considerably constrained in any attempt to modify their consumption patterns in order to minimize environmental damage. Because they believe that, as individuals, they cannot influence such matters, many consumers see little point in even trying to. Others appear to be unaware of any connection between their way of life and its environmental consequences.

The concept of psychographic segmentation provides a useful framework for examining trends in Green consumption patterns (O'Brien & Harris 1991). The method attempts to differentiate individuals on the basis of personality types that reflect personal motivation. For example, the Taylor Nelson Applied Futures model of the transformations taking place in British society suggests that three main groups of consumers can be identified by their attitudes, values, motivations, hopes and fears rather than by more familiar characteristics of social class, income and political viewpoint. This model has been cited by Porritt & Winner (1988), as a possible starting point for dissecting the process of the

Greening of consumers.

The main objective of the first of these groups, the Sustenance-Driven group, is to make ends meet. People in this group are characterized by relatively low levels of disposable income and consequently cannot really contemplate paying the premium prices attached to many "Green" products such as organically grown vegetables and stone-ground wholemeal bread. A second group, the Outer-Directeds, are presented as being preoccupied with striving for financial success, material trappings and social status. Any Green purchases made by Outer-Directeds are, "typically to make a statement about the individual rather than the state of the planet" (O'Brien & Harris 1991: 160). Finally, the Inner-Directeds are believed by Porritt & Winner (1988) to have the brightest Green potential as a consumer group. Inner-Directeds tend to have a holistic view of life, with a better-developed awareness of the environmental consequences of their behaviour. For them, individual development and self-expression is more important than acquiring material goods.

However the crude aggregation of consumers into these three groups is somewhat problematic. For example, "Inner-Directeds are by no means all Greens. Indeed, they are a pretty mixed bunch" (Porritt & Winner 1988: 173). This would appear to limit the applicability of such a categorization somewhat, especially given the point made above about the breadth of Green interpretations. As with retailers, consumers can choose from an array of consumption patterns that they perceive to be Green, often without really having to adjust their pattern of shopping behaviour very much, yet still managing to convince themselves that they have helped to save the planet.

In order to assess the extent to which the Greening of consumers has been occurring, a survey of 1,000 respondents was undertaken during 1991. The study was carried out in four towns in southern England: Cheltenham, Cirencester, Stroud and Swindon. For each town, a quota sample interview questionnaire survey of 250 consumers was completed in both central and out-of-centre shopping areas. The questionnaire was designed to explore the following: consumers' awareness of environmental problems; the extent to which purchasing behaviour had been modified as a result of heightened awareness of Green issues; the motivations driving any changes made to purchasing behaviour; and willingness to make further changes to consumption patterns. Just under 35 per cent of the sample revealed significant Green traits in their consumption behaviour. Of the remaining 65 per cent, some claimed to be completely uninterested in environmental matters, while others felt that although they might make occasional Green purchases, any individual contribution was too insignificant to make any real difference to environmental problems. A number expressed the view that, although sympathetic in principle, they were precluded from buying Green goods on a regular basis because of their higher prices. The relative incidence of the various responses is shown in Figure 12.1.

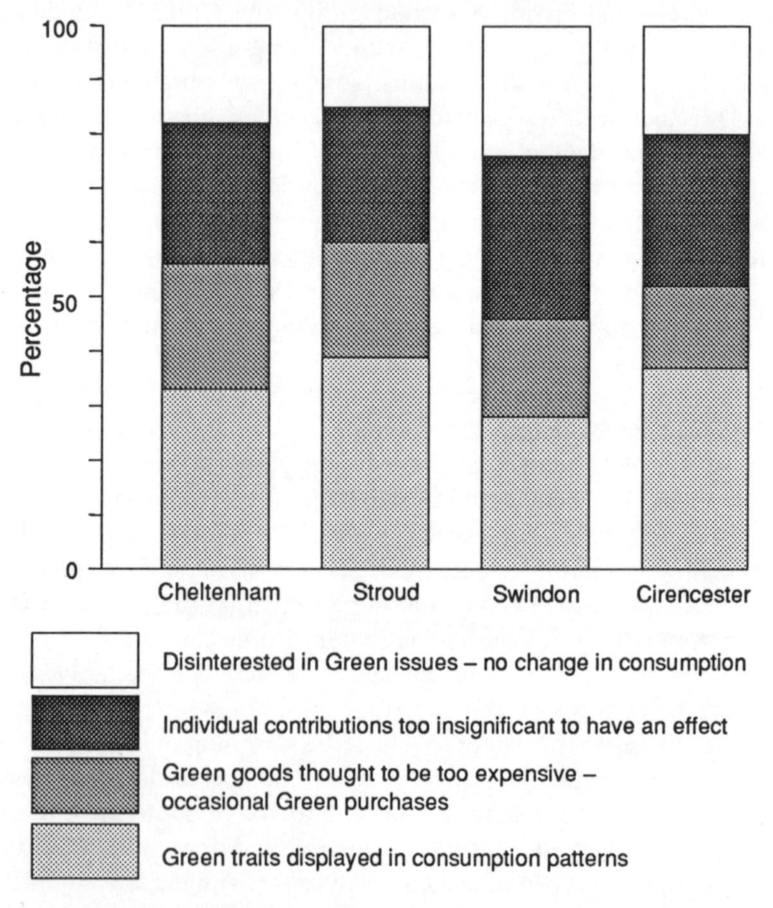

Figure 12.1 Percentage distribution of attitudes and consumption patterns in four English towns, 1991.

Though not necessarily indicative of a countrywide trend, owing to the sample size and locations selected for the survey, it is perhaps worth drawing attention to the fact that the figures for the smaller towns of Cirencester (1991 population 17,085) and Stroud (1991 population 22,923) reveal a greater presence of consumers displaying Green traits. The case of Stroud is particularly interesting when considered alongside the fact that the local Green Party candidates had a successful record of election to the local council during the late 1980s. The background to this local expression of Green awareness is inevitably complex, but the example does serve to point out that there may be significant spatial variability in Green markets and consumption patterns because of peculiar local circumstances.

Because of the nature of the sampling scheme adopted, information gathered from the questionnaire was not subjected to formal statistical analysis. Preliminary analysis of the survey returns suggested that a number of systematic patterns existed among the 35 per cent of respondents displaying Green consumption behaviour. A qualitative analytic approach was adopted with the intention of constructing a subjective classification of Green consumer types based on motivational behaviour patterns. The technique chosen was similar to that used by Allen (1983) for defining housing classes. Cross-tabulation of the survey responses permitted the identification and abstraction of five sets of Green consumer characteristics based on psychographic variables. By isolating the principal motivating reasons behind the respondents' purchasing behaviour and consumption, it became possible to create the following ideal type categories.

- *Committed Greens* (3 per cent of respondents) have the best developed philosophical and ideological understanding of what it means to be a Green consumer. Not only do they buy benign products but they also seek to reduce their consumption of material goods. Members of this group were the most enthusiastic recyclers and the most likely to be actively involved in Green politics.
- *Green "feel-goods"* (19 per cent of respondents) gain a glowing sense of satisfaction from feeling that they have done their bit for the environment. They are prepared to switch to selected Green products such as washing-up liquids, detergents, natural cosmetics, recycled paper and tissue products and unleaded petrol. Trips to supermarkets may well be combined with depositing empty bottles and jars at bottle banks. This group frequently becomes involved with short-term local environmental initiatives such as litter pick-ups and charity fund raising.
- *Conscience salvers* (7 per cent of respondents) are only willing to make a very limited adjustment to their patterns of consumption. Though frequently critical of the perceived under-performance of Green products, conscience salvers can gain a spurious sense of virtue from having changed, for example, to a new car fitted with a catalytic converter.
- *Self-concerneds* (6 per cent of respondents) are particularly interested

in a personal programme of healthy living, motivated among other things by a desire for increased longevity and the "body beautiful". Dietary awareness, exercise and fitness play a prominent part in their life styles. These motives have clear implications for particular forms of perceived Green consumption such as bottled mineral water, low-fat foods, wholefood, organically grown vegetables, mountain bikes, running shoes and natural cosmetics.

- *Rural-life-style seekers*. The 35 per cent of the sample for whom positive Green consumption traits could be identified only included the four groups discussed above. However, a frequently expressed secondary motivation driving purchasing behaviour suggests a fifth group of perceived Green consumption characteristics. This group's behaviour can be associated with the pursuit of a Green or rural life style and may be superimposed on the first four classes of Green consumer. They are well represented amongst the population that has been involved in the counter-urbanization process of the 1970s and 1980s (Champion & Watkins 1991).

Considerable variation is possible in the manifestation of these traits across the other four categories. For example, among committed Greens a frequently occurring theme was the attraction of living a simple, non-materialistic life in a tranquil rural environment. Moving away from this end of the spectrum of behaviour however, it is possible to identify elements of the "boutique ecologist".

The boutique ecologists, living in converted barns, renovated cottages and newly built "traditional, country-style homes" strive to create a manicured rusticity with painstaking attention to detail. Typical purchases might include craftsman-built kitchens, stripped-pine furniture, terracotta floor tiles, green waxed-cotton jackets and wood-burning stoves. None of these are in themselves Green products, but together they contribute to a romantic image of the pastoral idyll that is as satisfying for consumers as it is profitable for retailers. Once securely established in a rural environment, boutique ecologists often become extremely vocal in their opposition to subsequent developments that threaten to change its character.

In recognizing that the rural-life-style-seeker category cuts across the first four classes of Green purchasing behaviour, it is apparent that Green consumer types are not mutually exclusive. Hybrids are quite likely to occur given the many ways in which Green awareness can be interpreted. Green consumers are certainly not a homogeneous group and it may be quite possible for the same consumer to behave in very different ways regarding purchasing decisions in different Green product markets. Inclusion of the element of spatial variability into the distribution of Green consumer characteristics adds further complexity to the picture. Without doubt, the segmentation of Green consumers into distinct groupings is less clear cut than some approaches would have us believe (Gardner & Sheppard 1989, Peattie 1992).

Conclusions

There is no doubt that consumers and retailers are increasingly taking note of Green issues, but the available evidence suggests that these mean different things to different people. The term "Green" applied to consumption remains a chaotic conception cutting across a wide variety of interests and motivations on the part of all the actors. However, while there is clear evidence of an interest in environmental matters among consumers, there is much less certainty about its economic strength. It is not fully clear that there is sufficient commercial power among consumers to force the retailing system to supply only Green products. Given the diversity of Green consumers, there is still a major market for generic products offered at discounted prices alongside Green products. This situation is liable to remain the case for the foreseeable future given the current depressed state of the economy in the early 1990s, high levels of unemployment, the financial instability of sterling and the lack of consumer-led growth in the market.

Despite the indication that Green consumption may be commercially more limited than anticipated in the late 1980s, it is clear that people are concerned about the environment, even if this concern is rather vaguely expressed or leads to contradictory practices. The environmental views of children are particularly important and can have a significant influence on family-purchasing decisions, both now and in the future. Issues such as excessive packaging and recycling are easily turned into educational projects with clearly visible outcomes. Retailers may not wish to run the risk of being labelled as environmentally unfriendly by the next generation of consumers. Retailers and producers also have to take note of the environmental dimension of their activities, if only to protect their commercial image and market share. They are already accustomed to keeping a close watch on their competitors, and they will also need to monitor their rivals' environmental strategies. Failure to do so may have detrimental effects on their competitiveness. Likely outcomes of such a scenario include the continued growth of Green product and process innovations and an increasing use of Green imagery in promotional material and labelling.

In their rôle as voters, consumers have helped to establish the environment firmly on the political agenda. A range of environmental protection legislation has been enacted at both national and supranational levels, albeit with patchy, variable effects. There is every likelihood that the legislative framework will be extended and tightened, with pronounced implications for the Greening of manufacturers and retailers. Future developments in research and refinements in the scientific understanding of environmental problems will have implications for both retailers and consumers. Retailers will be faced with having to respond to a lengthening environmental agenda, while consumers will benefit from having access to more information on which to base both their political activity and their purchasing decisions.

To date, the responses of both retailers and consumers to Green issues only represent tentative steps although these may be the catalysts for future reform. Unlimited production and consumption is not sustainable but "the core of the Green message - that we must consume less - is being submerged under a wave of advertising and publications urging us to save the world by simply consuming better" (Irvine 1989: 88). However, breaking the mould will be difficult because the established practices of the economic system are more convenient for firms that have prospered and consumers that have benefited from them. For real progress to be made producers and retailers need to be cajoled and pressured into making and selling more benign products. But the bulk of consumers also need to be persuaded to tone down their materialistic excesses and hedonistic expectations. The process will be arduous because, as Mishan (1990: 35) notes, "the very characteristics of the age we live in as reflected in its basic aspirations and institutions, make it difficult for the peoples of the West, in their capacity as citizens, consumers and producers, consistently to rank environmental commitment before material considerations".

Time will tell whether there will be a radical change in the values and attitudes of the majority of consumers away from rampant materialism. Similarly, the extent to which retailers and producers will be able to forge a really meaningful form of "caring capitalism" is questionable. While it can be accepted that they *care* as people, capitalism has to come first, otherwise the rationale for the whole exercise falls apart.

CHAPTER THIRTEEN
Retail change: retrospect and prospect

Rosemary D. F. Bromley & Colin J. Thomas

In recent years radical change has occurred in the retail environment of most developed countries. This has affected both the business organization of retailing and its spatial expression at scales ranging from the international to the local. The activities of the largest retail firms have increasingly crossed international frontiers, national systems of distribution have become more sophisticated, and business concentration in the form of national and multinational corporations characterize retail operations.

At the same time, new types of retail facilities have emerged with a variety of impacts on the former system. New patterns of shopping behaviour have both precipitated and responded to these changes. Most consumers have become more mobile with the all-pervasive growth of car ownership, while the growth of affluence has resulted in demands for more and more specialized goods and services. It also appears that the situation of the less mobile, disadvantaged consumers has deteriorated as a result of these changes.

More explicitly, social issues have also come to the fore. The changing working conditions and hours of retail employees are, for example, creating a distinct divide between the relatively small numbers of executives and managers, and the large number of less-skilled workers. This has implications for manpower planning policy and for employees' social life. Wider changes in the working conditions of the population at large have also begun to highlight the potential leisure dimension of some aspects of the shopping experience for an increasingly significant proportion of the population. This is presenting a challenge to retailers to modify the whole nature of shopping activities, a challenge to which they are barely beginning to respond. Patterns of consumption are also being affected more directly by attitudes adopted to philosophical considerations beyond the immediate demands for goods and services. Foremost among these has been the environmental impact of consumption, while concerns relating to the health of the community and the economic and political implications of the origins of consumer goods have also been intermittently important. Such considerations have already presented important challenges to the retailing industry and this situation is likely to continue.

The analysis of these issues has formed the basis of this book, and the findings contribute towards a greater understanding of the changing nature of retailing for academics, and for those involved in retail planning, whether in the commercial or public service domains.

Changing retail organization

The first part of the book examined the trends towards economic concentration in national retail markets and the associated internationalization of retail business organization. Chapter 2 focused particularly on three principal trends in the development of international activity in retailing. Increasingly, retail firms were shown to exhibit international sourcing of goods; international operations, while still undertaken by only a minority of firms, were of steadily growing significance among the largest retailers; and few barriers appear to inhibit the international spread of marketing concepts, management ideals and technological innovations. Nevertheless, there are problems in the international transferability of retail concepts. Chapter 2 also raises this as a key issue in the internationalization process: the need for retailers to adapt to differing cultural contexts, even in retail environments that are superficially similar. However, despite a growing volume of research on the major trends, and a degree of affinity with internationalization in manufacturing, the complex operation of the forces at work requires far more detailed investigation if its implications are to become clear for the functioning of retailing in the future. Considerable variation seems likely to exist both between retail sectors and between countries.

A major determinant of such variations was identified in Chapter 3. The influence of the specifics of the domestic regulatory environments was shown to have far-reaching consequences for retail concentration in the grocery sector and for its ramifications in internationalization. The analysis focused initially upon a detailed case study of British grocery retailing, where in the 1980s the benign regulatory environment offered the opportunity for a small number of large firms to dominate the market, a trend completed by the early 1990s. This had fundamental implications for the spatial organization of grocery retailing, for the supplier–retailer relationships, for the nature of the workforce, and for the increasing use of computational information technology in all facets of business activities. The trend was also shown to have had major international repercussions. The need for opportunities for further expansion and the associated expectations of the financial markets were shown to result in capital switching involving international expansion programmes by the largest firms. The expansion of Sainsbury's into the USA and of Tesco into France provide clear demonstrations of this trend. Likewise, the partial abandonment by the market leaders of small store operations in the traditional shopping centres, combined with the attractions of the high profit margins achieved by British grocery retailers, offered the opportunity for the penetration of this market niche by cut-price retailers. The British company Kwik Save took early advantage of this opportunity with considerable success, but this market has also proved attractive to specialist European competition in recent years.

National and international competitive implications of the changing economic characteristics of retail business organization are evident from these analyses. Given the continuing importance of the trends towards retail concentration and internationalization in contemporary developed economies, there is a need for additional case studies and for the development of an encompassing theoretical framework within which to investigate the trends. Therefore, further analyses of this type, for both practical and theoretical purposes, are likely to feature prominently in the research agenda concerned with the processes of retail change.

The new retail phenomena

Chapter 4 provided additional insights into the international dimension of retail change, while providing a bridge between the economic and spatial analysis of retail innovation. The expectation of an international convergence of the nature of the new retail phenomena throughout Europe was challenged. Despite an underlying similarity of the trends between countries, it is apparent that individual variations in economic, social and institutional environments result in important differences in the characteristics of retail changes. This has direct implications for the future commercial opportunities for international expansion programmes.

A paucity of good-quality comparative data, however, continues to be an obstacle to the refinement of this type of international analysis. Yet a tentative European typology of planned shopping centres was postulated that provides at least the basis for a better understanding of the complex nature of the international variations in the physical expression of retail change. A recent exploratory attempt by Brown (1990) to develop a model combining contemporary retail form and function as a framework for the spatial analysis of retailing also demonstrates the complexity of the existing situation. The locational and compositional complexity of new planned shopping centres in Europe incorporated in the typology presented in Chapter 4 provides an additional international perspective. This serves to highlight further the need for a comprehensive conceptual model of retail structure to assist contemporary retail geographical analysis. Clearly, Johnson's (1991: 260) view that "We may have to find concepts other than the hierarchy to explain how parts of the retail puzzle fit in a spatial sense" awaits further clarification. Additional research effort could be usefully applied in this direction.

Planning implications of retail decentralization

The mobile shopper

A number of planning implications emerge from the analyses of the process and impact of retail change in the urban region. Common to most of the chapters in parts 2 and 3 of the book is the observation that car usage for shopping trips has popular appeal because of the perceived advantages of convenience associated with time and route flexibility. The contemporary political climate also views a vigorous car-manufacturing industry as a vital ingredient of national economic revival. Unless environmental concerns force a shift of emphasis back to public transport, it seems a foregone conclusion that car ownership will continue to increase steadily. This is likely to enhance further the attractions of the new car-oriented facilities. Thus, even if nothing else changes, a slow diversion of trade from traditional to modern shopping facilities is likely to continue, implying a continuing erosion of the older shopping centres and a deterioration of the shopping opportunities most easily accessible to the smaller residual section of the community that does not own cars.

Decentralization and planning constraint

However, a number of other important issues remain to be resolved. Central among these is the future relationship between the economic and social forces promoting further retail decentralization and the degree to which these will be constrained by the application of planning controls. Despite the relaxation of restrictive planning legislation on the full range of new retail forms throughout the 1980s and early 1990s, a *laissez faire* situation on the US model does not exist in the UK. Central government caution combined with a degree of ambiguity persists, particularly with regard to the development of regional shopping centres (DOE 1988, 1992a, 1992b). In addition, local government for the most part continues to urge restraint on further large-scale decentralization, and to stress the economic and social value of a strong city centre and the need to continue to serve the less mobile sections of the community.

The future of the city centre

Chapter 5 provided a graphic illustration of alternative and diametrically opposite city-centre experiences by focusing on the changes in the regional metropolitan centres of Cardiff, UK and Charlotte, USA. Fundamental differences in the nature and scale of suburbanization; the planning environment; the associated attitudes of financiers, developers and retailers to the prospects of the city centres; and the influence of a number of more localized factors, all combined to present either opportunities or obstacles to development. Clearly, the details of the process of change are complex. Nevertheless, the alternatives stood in stark contrast. At one extreme, unfettered decentralization is clearly capable of precipitating a spiral decline in the traditional centres, while a balanced policy based upon a significant element of constraint can offer a context within which the retention of a vibrant city centre is possible.

Chapter 6 developed this theme further, focusing more directly upon the likelihood of the negative scenario for the British city. The traditional ideas of CBD dominance now require serious qualification and reformulation. The established CBDs are failing to attract, and in fact are losing outlets and sales in significant growth sectors of retailing, such as furnishings, electrical goods, garden centres and DIY. The commercial weakening of the city centre in the larger conurbations is portrayed as a process that is well advanced as a consequence of the decline of the inner-city economy, the growing problems of traffic congestion, and, in particular, the negative impacts of the new out-of-centre retail developments. The emerging trend towards the decentralization of offices to suburban sites, and the likely continued decline in the accessibility of city centres because of traffic-management policies are liable to further exacerbate the problem. This view serves to underline the urgency of the situation. For some areas the future of the traditional shopping centres is not necessarily seen as dependent upon the formulation of a national retail planning policy. Instead, progressive decline may already be an emerging reality.

The impact of out-of-centre shopping

The various impacts of the process of retail change were the focus of Chapter 7. Current evidence suggests that the degree, nature and location of the newer retail forms are closely related to their effect on traditional centres. In addition, further substantial retail decentralization, incorporating increasingly specialized functions, appears capable of increasing the attractions of out-of-centre facilities for the more affluent mobile shopper. The already dominant rôle of the superstores for grocery retailing and the recent experiences of the regional shopping centres, provide ample evidence of this likelihood. Such an outcome is capable of initiating a spiral decline in the fortunes of the city centres and of the other smaller traditional centres. The creation of a situation closely akin to that in the USA cannot be ruled out.

By contrast, a degree of planned decentralization consistent with a complementarity between the old and the new appears capable of retaining substantial elements of the older system, while accommodating the new demands for retail innovation. The experience of the intermittently restrictive planning environment of the past 25 years is, at least, indicative of such an eventuality. In recent years many city centres have been subject to "compaction", while a weakening of the base of the other traditional centres has been suggested. Nonetheless, redevelopment and refurbishment of older centres of all sizes has been a common feature, while substantial decline has been confined largely to situations where attractive new facilities have been developed in relatively close proximity to older centres. In effect, the absence of widespread dire consequences of retail decentralization on the traditional centres before 1993 suggests the continued possibility of a degree of accommodation between the old and new. In view of the cautious attitude of central government in recent years, even within the context of an increasingly free-market orientation, a policy that attempts to incorporate the advantages of the traditional system while providing for retail innovation appears to be the most likely outcome.

However, the current absence of a clear and unambiguous national retail planning policy is likely to undermine confidence in the potential for future re-investment in the traditional centres. In these circumstances, a gradual erosion is liable to continue by default in the manner postulated in Chapter 6. The evidence, therefore, is indicative of the urgent need for firm policy guidelines, when viewed from the perspective of the traditional centres. This guidance is most desperately needed where the likely impacts of new shopping developments extend far beyond local government boundaries. Currently, the retail strategy for one district can be undermined by permitted developments in another, which is especially devastating where attempts at regenerating shopping provision in a depressed area are destroyed by policies elsewhere.

Revitalizing the city centre

Even if the planning issue is resolved in favour of a policy of accommodation, the future health of the city centre and other traditional centres will involve additional issues. Foremost is the question of accessibility, since existing access, traffic congestion and parking problems are already poised to exacerbate the situation. Evidence reviewed in this book suggests that improvements in accessibility will need to be given as much attention as the enhancement of the shopping attractions of the traditional facilities if the latter are to continue to flourish. Thus, improved transport and car-parking strategies, in particular the provision of high-turnover shopping car parks, will need to be closely integrated with retail planning policies. Within this context, public-transport planning is a potentially vital ingredient. Improved public transport is capable of enhancing the status of city centres, the smaller traditional centres and even the new facilities at nodal locations in the urban fabric, while also offering the opportunity to provide for the needs of consumers who do not have access to cars.

Of related significance is the issue of "management and promotion". Evidence from North America suggests that the co-ordination of the activities of city-centre interests in the public and private sectors has been important in the successful examples of retail revitalization in the declining centres. Such activities are still in their infancy in the UK, and the issues of leadership and finance remain to be resolved. Nevertheless, the limited available evidence suggests the potential value of this approach in supporting the status of traditional shopping centres.

Planning for shopper security

Also relevant is the question of shopper security in the city centre explored in Chapter 8. The often complex morphological structures of city centres resulting from organic and planned processes of development have multiplied the opportunities for crimes against people, property and cars. Fear of crime has consequently emerged in recent years as a contribution to the malaise of the city centre which, if left unchecked, could become an important component of the tendency towards a spiral of decline. Planners need to be aware of the wide array of urban-design devices that are together capable of reducing the fears of the public to manageable levels. Clearly, urban design cannot resolve the wider problem of crime in society. Yet, if the city centres are to compete on an equal footing with the newer forms of shopping centres, much greater attention needs to be paid to this issue. In the absence of such a reaction, some expensive multi-storey car parks as well as the less-frequented shopping streets could turn into commercial "no go" areas, and the tendency for city centres to resemble "shuttered fortresses" shortly after the closure of the principal shops can only accentuate the process of decline.

The evidence presented in this book stresses that if security initiatives are to succeed, whether in the city centre or in the new retail areas, a subtle approach will be required. Hopkins (1991), for example, highlights the inherent contradiction incorporated in design solutions to the problems posed by crime and vandalism in the shopping environment. Retailers and property developers are primarily concerned to protect their property, but at the same time they are equally concerned to provide a safe, attractive and inviting shopping experience for the public. Private-property protection and ease of public access are potentially conflicting aims. If, for example, shopping-centre design is focused on the needs of private-property protection, public access for informal social interaction or window-shopping outside shopping hours might well be reduced by closing some, or even all, pedestrian routes. Similarly, security devices such as window shutters are likely to reduce the general attractions of the shopping environment. A graphic example has been outlined for Milton Keynes (Francis 1991). The shopping centre is contained largely within a single "building", which is normally closed outside shopping hours. Consequently, for much of the time the centre is not accessible, but presents a barrier to pedestrians and necessitates circuitous pedestrian journeys, incorporating potentially adverse safety implications. Clearly such a situation is unlikely to promote the commercial attractions of the centre and to contribute to its positive image as a focus of community identity. Such considerations are becoming an all too common aspect of both traditional city centres and the newer forms of planned shopping centre in the UK and Europe (Poole & Donovan 1992).

Issues of social concern

The book incorporates an emphasis on several issues of social concern; the growing problem of crime in the shopping centre considered above is one such. Much of the discussion relating to the future of the city centre and the impact of the new retail facilities has widespread social implications, in particular for the disadvantaged consumer. Similarly, changes in the working week have unattractive implications for retail employees, who may be pressured into working unsocial hours.

The disadvantaged consumer

The existence of disadvantaged consumers who are constrained in their access to a wide variety of stores is now widely recognized, despite the fact that the topic is poorly researched (DOE 1992a, Bromley & Thomas 1993). Chapter 9 highlights the problems of the different categories of disadvantaged, and outlines solutions. Most of the policies that have been implemented have been partial and directed only at the alleviation of particular problems for special groups, such as access for the disabled. Unfortunately, a principal difficulty in tackling the problems of disadvantaged consumers is the limited power of intervention of the public sector. Two obvious strategies are either to encourage stores to locate in areas easily accessible to disadvantaged consumers, or to provide cheap and efficient public transport to the stores that already exist. In both cases the stores in question would need to provide good value and variety to remove "disadvantage". A third strategy is the provision of electronic home shopping, with terminals located in the home and goods delivered at a competitive price. This solution is currently too expensive to be economic without substantial subsidization. In addition, it has the disadvantage of removing the invaluable social-contact dimension of shopping. Further research is required both to further understanding of the needs of differing groups of disadvantaged and to identify the most appropriate policy measures.

Working at weekends

Consumer demand for longer shopping hours has resulted in extended retail provision and associated pressure on the retail workforce to work unsocial hours, including Saturdays and Sundays. This pressure has considerable social implications for shop workers, over half of whom are women. Chapter 10 focuses particularly on the attitudes of retail employees in the UK to working unsocial hours; 40 per cent, for example, preferred not to work on Sundays. Sixty per cent affirmed that if they had to work on Sundays it would reduce the time they spent with their families. Unfortunately, the poorly qualified retail employee is particularly vulnerable to employer pressure to work such unsocial hours. Although a legislative solution might not be appropriate, it is vital that those planning the extended availability of retail provision should consider the welfare of the shop worker.

Social trends in retailing

Other social issues raised in the book relate to trends rather than concerns, but both present challenges and opportunities to the retailers and planners.

Green issues

The Greening of shopping, which developed as a major trend in the 1980s, was examined in Chapter 12. "Green" has been used in a plethora of ways and the freedom of interpretation has allowed manufacturers and retailers considerable licence to declare their behaviour to be Green and capitalize on the consequences. The motives lying behind retailers' Green responses are varied, and the question emerges whether, under capitalism, retailers can ever effectively reconcile the quest for profit with a commitment to Green issues. On the consumer side, there is widespread evidence of an interest in environmental matters, but there is less certainty about its economic strength. A survey of consumers revealed that only 35 per cent displayed positive Green consumption traits. The commercial power among consumers for the Greening of shopping is likely to be weaker in the 1990s, especially if the adverse economic climate continues. Nevertheless, progress to date may be the base for future more dramatic reform in the future.

Leisure shopping

The growing rôle of shopping as a leisure activity has emerged as a social trend accompanying the marked growth in leisure time. Chapter 10 explored the characteristics of leisure shopping and identified consumer frustration with the current predictability of the retail environment. Those involved in retail planning and development should respond to new leisure-shopping behaviour. The planned combination leisure–retail centres, incorporating food courts offering the opportunity for relaxation and refreshment, are one strategy designed to enhance the shopping experience. Alternatively, the entertainment-based response can be adopted on a variety of scales in most shopping environments. Festival shopping has been a strategy to revive or maintain the attraction of many shopping centres, with the same aims of improving the shopping environment and enhancing retail profits. Retail providers should also strive to design relaxing and attractive surroundings and assist in re-establishing the uniqueness of the shopping environment.

Concluding comments

The overriding impression created by the foregoing analyses of retail change is the sheer dynamism of the phenomena. Retail organizations are continuing to grow in scale and variety, and to extend their operations across international frontiers with varying degrees of success. The variety of new retail forms continues to expand, with varying effects on the previously dominant systems. Information technology is generating innovation in business administration, the structure of the distribution chain, stock control, and point of sale procedures in ways unimagined even 10 years ago. The future of the city centre in the UK and Europe is still in doubt, and is likely to reflect the nature of government controls, the future status of public transport policies, and their influence on the climate of investment. Governmental attitudes to the degree and nature of further retail decentralization are still unexplicit. Social trends associated with retail change also have social consequences; the plight of the disadvantaged consumer could get worse, as might the working conditions of retail employees. Shopper safety and the security of shop property are rapidly growing issues in the traditional and new shopping environments alike, and are presenting major problems for both planners and the police. On a brighter note, the retail industry is likely to continue to enhance the shopping experience by responding to consumer demands for a stronger leisure orientation. The future significance of the "Green" dimension, however, remains a major imponderable for manufacturers, retailers and consumers.

Clearly, retail change is an ongoing process and many aspects of the contemporary issues identified in this book remain to be resolved. The dynamism of the phenomena is such that retail change will continue to present new challenges to retail analysts undertaking investigations at the wide range of geographic scales and academic disciplines contained in this book. The retail revolution is far from reaching an end.

REFERENCES

Access for All 1990. *Access for all: information pack*. Birmingham: Access for All.

Adams, C. (ed.) 1992. *Guide to the antique shops of Britain 1992*. Woodbridge, Suffolk: Antique Collectors Club.

Aharoni, Y. 1966. *The foreign investment decision process*. Cambridge, Mass.: Harvard University Press.

Alexander, I. & J. A. Dawson 1979. Employment in retailing: a case study of employment in suburban shopping centres. *Geoforum* **10**, 407–25.

Alexander, N. 1990. Retailers and international markets: motives for expansion. *International Marketing Review* 7(4), 75–85.

Allen, J. 1983. Property relations and landlordism - a realist approach. *Environment and Planning* D **1**, 191–203.

Anderson, V. 1991. *Alternative economic indicators*. London: Routledge.

Atkins, S. 1989. *Critical paths: designing for secure travel*. London: The Design Council.

Audit Commission 1989. *Urban regeneration and economic development: the local government dimension*. London: HMSO.

Baerwald, T. 1989. Changing sales patterns in major American metropolises, 1963–1982. *Urban Geography* **10**, 355–74.

Balchin, P. N. & G. H. Bull 1987. *Regional and urban economics*. London: Harper & Row.

Beckerman, W. 1992. Economic growth and the environment: Whose growth? Whose environment? *World Development* **20**, 481–96.

Bennison, D. 1985. Domestic viewdata services in Britain: past experience, present status, and future potential. *Environment and Planning* B **12**, 151–64.

Bernard Thorpe & Partners 1985. *Retail warehouse parks: an approach to planned development*. London: Bernard Thorpe & Partners.

Bilkey, W. & E. Nes 1982. Country of origin effects on product evaluations. *Journal of International Business Studies* **13**(1), 89–99.

Birmingham City Council 1990. *Multi-storey car park design survey*. Birmingham: Birmingham City Council.

Blacksell, M. 1991. Leisure, recreation and environment. In *The changing geography of the UK*, 2nd edn, R. J. Johnston & V. Gardiner (eds), 362–81. London: Routledge.

Body Shop 1991. *Body Shop Broadsheet* (August). Littlehampton: The Body Shop.

Borchert, J. G. 1988. Planning for retail change in the Netherlands. *Built Environment* **14**, 22–37.

Boseley, S. 1992. Cash for homeless wasted by dogma. *The Guardian* 2 November, 7.

Bosquet, M. 1977. *Capitalism in crisis and everyday life*. Brighton: Harvester.

Bowlby, S. 1979. Accessibility, mobility and shopping provision. In *Resources in planning*, A. Kirby & B. Goodall (eds), 293–323. Oxford: Pergamon.

Bowlby, S. 1984. Planning for women to shop in post-war Britain. *Environment and*

Planning D **2**, 179–99.

Bowlby, S. 1985. Shoppers' needs. *Town and Country Planning* **54**, 219–22.

Bowlby, S. 1988. From corner shop to hypermarket: women and food retailing. In *Women in cities, gender and the urban environment*, J. Little, L. Peake, P. Richardson (eds), 61–83. London: Macmillan.

Bowlby, S. R. & J. Foord 1989. The changing organization of retailer–supplier relationships: locational implications. Paper presented at Regional Science Association 29th European Congress, Cambridge; mimeo available from Department of Geography, University of Reading.

BPC (British Productivity Council) 1953. *Anglo-American productivity team report on retailing*. London: British Productivity Council.

Breheny, M. J. 1990. Visions for the South, *Royal Society for the Encouragement of Arts, Manufactures and Commerce, Journal* **5401**, 691–701.

Broehl, W. G. (ed.) 1968. *United States business performance abroad: the case study of International Basic Economy Corporation*. Washington, DC: National Planning Association.

Bromley, R. D. F. & C. J. Thomas 1988. Retail parks: spatial and functional integration of retail units in the Swansea Enterprise Zone. *Institute of British Geographers, Transactions* **13**, 4–18.

Bromley, R. D. F. & C. J. Thomas 1989. Clustering advantages of out-of-town stores. *International Journal of Retailing* **4**(3), 41–59.

Bromley, R. D. F. & C. J. Thomas 1990. *Household shopping survey in Swansea*. Swansea: Swansea City Council.

Bromley, R. D. F. & C. J. Thomas 1993. The retail revolution, the carless shopper and disadvantage. *Institute of British Geographers, Transactions* **18**(2).

Broom, D. 1992. Planning rules are blamed for threat to countryside. *The Times* (8 August), 6.

Brown, S. 1987. The complex model of city centre retailing: an historical application. *Institute of British Geographers, Transactions* **12**, 4–18.

Brown, S. 1989. *Retail warehouse parks*. Harlow, England: Longman.

Brown, S. 1990. Retail location: the post hierarchical challenge. *International Review of Retail, Distribution and Consumer Research* **1**(3), 367–81.

Buchanan, C. & Partners 1968. *Cardiff development and transportation study*. Cardiff: Cardiff City Council.

Buckley, P. J. 1983. New theories of international business. In *The growth of international business*, M. Casson (ed.). London: Allen & Unwin.

Buckwalter, D. W. 1989. Effects of competition on the pattern of retail districts in the Chattanooga, Tennessee, metropolitan area. *Southeastern Geographer* **29**, 26–41.

Buckwalter, J. 1987. Securing shopping centres for inner cities. *Urban Land* (April), 22–5.

Burke, T. 1986. *The likely impact of deregulation of Sunday trading on consumption and employment*. London: Faculty of Social Science and Business Studies, Polytechnic of Central London.

Burt, S. 1990. *The internationalization of European retailers*. Working Paper 9005, Institute for Retail Studies, University of Stirling.

Burt, S. L. 1989. Trends and management issues in European retailing. *International Journal of Retailing* **4**(4), 3–97.

Burt, S. L. 1991. Trends in the internationalization of grocery retailing: the European experience. *International Review of Retail Distribution and Consumer Research* **1**, 487–515.

REFERENCES

Burt, S. L. & J. A. Dawson 1989. L'internazional izzazione del commercio al dettaglio Inglese. *Commercio* **35**, 137–57.

Burt, S. L. & J. A. Dawson 1991. *The impact of new technology and new payment systems on commercial distribution in the European Community*. DG XXIII, Series Studies, Commerce and Distribution 17. Brussels: Commission of the European Communities.

Butler, R. W. 1991. West Edmonton Mall as a tourist attraction. *Canadian Geographer* **35**, 287–95.

Cameron, G. C. (ed.) 1980. *The future of British conurbations: policies and prescription for change*. Harlow, England: Longman.

Carey, R. J. 1988. American downtowns: past and present attempts at revitalization, *Built Environment* **14**, 46–70.

Carol, H. 1962. The hierarchy of central functions within the city: principles developed in a study of Zurich, Switzerland. In *The IGU Symposium in Urban Geography, Lund 1960*, K. Norberg (ed.), 555–76. Lund: Gleerup.

Carr, J. 1990. The social aspects of shopping: pleasure or chore? the consumer perspective. *Royal Society of Arts Journal* **138**, 189–97.

Carson, D. 1967. *International marketing*. New York: John Wiley.

Carter, H. & G. Rowley 1966. The morphology of the central business district of Cardiff. *Institute of British Geographers, Transactions* **38**, 119–34.

Carter, H. 1981. *The study of urban geography*. 3rd edn. London: Edward Arnold.

CEC (Commission of the European Commission) 1990. *Green paper on the urban environment*. Brussels: Commission to the Council and Parliament.

CEC (Commission of the European Communities) 1991. *Development on the labour market in the Community: results of a survey covering employers and employees*. European Economy No. 47. Brussels: Directorate General for Economic and Financial Affairs.

Champion, A. G. (ed.) 1989. *Counterurbanization: the changing pace and nature of population deconcentration*. London: Edward Arnold.

Champion, A. G. 1992. Urban and regional demographic trends in the developed world. *Urban Studies* **29**, 461–82.

Champion, A. G. & A. R. Townsend 1990. *Contemporary Britain: a geographical perspective*. London: Edward Arnold.

Champion, A. G. & C. Watkins (eds) 1991. *People in the countryside: studies of social change in rural Britain*. London: Paul Chapman.

Citizens Crime Commission of New York City 1985. *Downtown safety, security and economic development*. New York: Downtown Research and Development Centre.

City of Cardiff 1991. *Land use and floorspace survey 1990: Cardiff city centre*. Cardiff: Cardiff City Council.

Clark, G. L. 1989. Remaking the map of corporate capitalism: the arbitrage economy of the 1990s. *Environment and Planning* A **21**, 997–1000.

Clark, G. L. 1993. Strategy and structure: corporate restructuring and the scope and characteristics of sunk costs. *Environment and Planning* A **25**.

Clarke-Hill, C. M. & T. M. Robinson 1992. Co-operation as a competitive strategy in European retailing. *European Business and Economic Development* **1**(2), 1–6.

CNRS (Centre Nationale de Recherches Scientifique) 1991. Fonctions urbaines et avenir du commerce de centre ville, *Première Recontre d'Urbanisme Commercial CNRS/SÉGÉCÉ*. Paris: CNRS.

Collins, R. G. et al. 1991. *America's downtowns*. Washington, DC: The Preservation

Press.
COMEDIA (in association with Gulbenkian Foundation) 1991. *Out of hours*. London: Comedia.
Conrads, U. 1970. *Programmes and manifestos on twentieth century architecture*. London: Lund Humphries.
Corporate Intelligence Group 1990. *International retailers in Europe. Oxford: Oxford Institute of Retail Management.*
Corporate Intelligence Group 1992a. *Cross-border retailing in Europe*. London: Corporate Intelligence Group.
Corporate Intelligence Group 1992b. *European directory of shopping centres* (2 vols). London: Corporate Intelligence Group.
County NatWest WoodMac 1991a. *Kwik Save Group: the choice is obvious*. London: County NatWest Securities.
County NatWest WoodMac 1991b. *Focus on European food retailing*. London: County NatWest Securities.
Crewe, L. & E. Davenport 1992. The puppet show: changing buyer–supplier relationships within clothing retailing. *Institute of British Geographers, Transactions* **17**, 183–97.
Cullingworth, J. B. 1989. Planning problems and policies. *Cities* **6**, 151–5.
Cunningham, H. 1980. *Leisure in the industrial revolution*. London: Croom Helm.

Davies, G. & J. Bell 1991. The grocery shopper - is He different? *International Journal of Retail and Distribution Management* **19**(1), 25–8.
Davies, K., C. Gilligan, C. Sutton 1985. Structural changes in grocery retailing: the implications for concentration. *International Journal of Physical Distribution and Materials Management* **15**, 3–48.
Davies, K., C. Gilligan, C. Sutton 1986. The development of own-label product strategies in grocery and DIY retailing in the United Kingdom. *International Journal of Retailing* **1**, 6–19.
Davies, K. & L. Sparks 1989. The development of superstore retailing in Great Britain 1960–1986: results from a new database. *Institute of British Geographers, Transactions* **14**, 74–89.
Davies, R. L. 1972. Structural models of retail distribution: analogies with settlement and land use theories. *Institute of British Geographers, Transactions* **57**, 59–82.
Davies, R. L. 1973. *The retail pattern of the central area of Coventry*. Institute of British Geographers Occasional Publications 1, 1–42.
Davies, R. L. 1984. *Retail and commercial planning*. London: Croom Helm.
Davies, R. L. 1985. The Gateshead Shopping and Information Service. *Environment and Planning* B **12**, 209–20.
Davies, R. L. 1986. Retail planning in disarray. *The Planner* **72**, 20–2.
Davies, R. L. 1991. *Retail change of Tyneside: the impact of the MetroCentre*. End of award report D00232171. Swindon: ESRC.
Davies, R. L. & A. G. Champion 1980. *Social inequality in shopping opportunities: how the private sector can respond*. Newcastle: The University and Tesco.
Davies, R. L. & A. G. Champion (eds) 1983. *The future for the city centre*. London: Academic Press.
Davies, R. & E. Howard 1988. Issues in retail planning within the United Kingdom. *Built Environment* **14**, 7–21.
Davies, R. L. & E. B. Howard 1989. The Metro Centre experience and the

prospects for Meadowhall. *Retail and Distribution Management* **17**(3), 8–12.
Davies, R. L. & J. Reynolds 1988. *The development of teleshopping and teleservices*. Harlow, England: Longman.
Davies, S. C. 1982. *Designing effective pedestrian improvements in business districts*. Chicago: APA publications.
Dawson, J. A. 1983. *Shopping centre development*. Harlow, England: Longman.
Dawson, J. A. 1985. *The costs and returns of Sunday trading*. Working Paper 8502, Institute for Retail Studies, University of Stirling.
Dawson, J. A. 1978. International retailers. *Geographical Magazine* **51**, 248–9.
Dawson, J. A. 1991. *Le commerce de détail Européen*. Paris: Presses du Management.
Dawson, J. A. & S. A. Shaw 1992. *Inter-firm alliances in the retail sector: evolutionary, strategic and tactical issues in their creation and management*. Working Paper, 92/7. Edinburgh: University of Edinburgh, Department of Business Studies.
Dawson, J. A., D. Gransby, R. Schiller 1988. The changing High Street. *Geographical Journal* **154**, 1–22.
Dawson, J. A. & J. D. Lord (eds) 1985. *Shopping centre development: policies and prospects*. Beckenham: Croom Helm.
Dawson, J. A. & S. A. Shaw 1990. The changing character of retailer–supplier relationships. In *Retail distribution management*, J. Fernie (ed.), 19–39. London: Kogan Page.
Dawson, J. A. & L. Sparks 1987. Issues for the planning of retailing in Scotland. *Journal of Scottish Planning Law and Practice* **18**, 38–40.
Department of Transport 1988. *National travel survey: 1985/6 Report, Part I: an analysis of personal travel*. London: HMSO.
DHI (Deutsches HandelsInstitut) 1990. *Shopping-center report*. Köln: ISB-Institut fur Selbstbedienung und Warenwirtschaft GmbH.
Diamond, D. R. 1991. Managing urban change: the case of the British inner city. In *Global change and challenge: geography for the 1990s*, R. Bennett & R. Estall (eds), 217–41. London: Routledge.
Dicken, P. 1986. *Global shift: industrial change in a turbulent world*. London: Harper & Row.
Dobson, A. 1990. *Green political thought: an introduction*. London: HarperCollins Academic.
DOE (Department of the Environment and Welsh Office) 1988. *Major retail development*. Planning Policy Guidance 6. London: HMSO.
DOE (Department of the Environment/BDB Planning/Oxford Institute of Retail Management) 1992a. *The effects of major out-of-town retail development*. London: HMSO.
DOE (Department of the Environment and Welsh Office) 1992b. *Revised PPG6 – Town centres and retail development*. Consultation draft. London: HMSO.
DRDC (Downtown Research and Development Center) 1988. *How downtowns organize for results*. New York: DRDC.
Drucker, P. 1958. Marketing and economic development. *Journal of Marketing* **22**, 252–9.
DTEDC (Distributive Trades Economic Development Committee) 1973. *The distributive trades in the common market*. London: HMSO.
DTEDC (Distributive Trades Economic Development Committee) 1988. *The future of the High Street*. London: HMSO.
Ducatel, K. & N. Blomley 1990. Rethinking retail capital. *International Journal of Urban and Regional Research* **14**, 207–27.

Dunning, J. H. 1981. *International production and the multinational enterprise*. London: Allen & Unwin.

Dunning, J. H. & G. Norman 1985. Intra-industry production as a form of international economic involvement: an exploratory analysis. In *Multinationals as mutual invaders: intra-industry direct foreign investment*, A. Erdilek (ed.) London: Croom Helm.

EC (European Commission) 1991. *Towards a single market in distribution*. Communication of European Commission, COM91 41. Brussels: European Commission.

Edvardsson, B., L. Edvinsson, H. Nystrom 1992. Internationalization in service companies. *Service Industries Journal* **13**(1), 80–97.

Ehrlich, P., A. H. Ehrlich, J. P. Holdren 1973. *Human ecology: problems and solutions*. San Francisco: W. H. Freeman.

Ekins, P. (ed.) 1986. *The living economy*. London: Routledge & Kegan Paul.

Elkington, J. & J. Hailes 1988. *The Green consumer guide*. London: Gollancz.

Employment Gazette 1992. Labour market commentary. *Employment Gazette* July, **100**(7).

Eroglu, S. 1992. The internationalization process of franchise systems: a conceptual model. *International Marketing Review* **9**(5), 19–30.

Eversley, D. 1990. Inequality at the spatial level: tasks for planners. *The Planner* **76**(12), 13–18.

Fagan, M. L. 1991. A guide to global sourcing. *Journal of Business Strategy* **12**(2), 21–5.

Fairbairn, K. J. 1991. West Edmonton Mall: entrepreneurial innovation and consumer response. *Canadian Geographer* **35**, 261–8.

Feinstein, S. S. 1991. Promoting economic development: urban planning in the United States and Great Britain. *American Planning Association, Journal* **57**, 22–33.

Ferris, J. S. 1991. On the economics of regulated early closing hours: some evidence from Canada. *Applied Economics* **23**, 1393–400.

Ferry, J. W. 1960. *A history of the department store*. New York: Macmillan.

Fieller, L. & J. Peters 1991. *Planning principles for Chicago's Central Avenue*. Chicago: City of Chicago.

Filser, M. 1990. L'enjeu stratégique du marché unique pour les firmes de distribution. Working Paper, 9005. Dijon: CREGO, Université de Bourgogne.

Flynn, A. & T. Marsden 1992. Food regulation in a period of agricultural retreat: the British experience. *Geoforum* **23**, 85–93.

Foord, J., S. Bowlby, C. Tillsley 1992. Changing relations in the retail-shopper chain. *International Journal of Retail and Distribution Management* **20**(5), 23–30.

Ford, R. 1991. Some perspectives on retailing in the 1990s. *International Journal of Retail and Distribution Management* **19**(5), 17–21.

Foster, A. 1981. Push button shopping – when will it happen? *Retail and Distribution Management* **9**(1), 29–32.

Francis, A. 1991. Private rights in the city centre. *Town and Country Planning* **60**, 302–3.

Freiden, B. J. & L. B. Sagalyn 1989. *Downtown Inc.: how America rebuilds cities*. Cambridge, Mass: MIT Press.

Fyson, A. 1991. Reality returns to retailing. *The Planner* **77**(1), 3.

REFERENCES

Gaedeke, R. 1973. Consumer attitudes to products "made in" developing countries. *Journal of Retailing* **49**(1), 13–24.

Gapps, S. 1987. Global marketing. *Management Horizon Retail Focus*, Summer, 3–78.

Gardiner, C. & J. Sheppard 1989. *Consuming passion: the rise of retail culture*. London: Unwin Hyman.

Garner, B. J. 1966. *The internal structure of retail nucleations*. Evanston, Illinois: Northwestern University.

Gascoigne, R. 1992. The entry of European limited-line discounters into the UK. Paper presented at Institute of British Geographers Annual Conference, Swansea; mimeo available from Department of Geography, University of Southampton.

Gascoigne, R. 1993. The restructuring of British food retailing: the entry of European limited-line discounters into the UK. PhD dissertation, Department of Geography, University of Southampton.

Gayler, H. 1989a. *Retail innovation in Britain: the problems of out-of-town shopping centre development*. Norwich: Geo Books.

Gayler, H. J. 1989b. The retail revolution in Britain. *Town and Country Planning* **58**, 277–80.

Getis, A. & J. M. Getis 1968. Retail store spatial affinities. *Urban Studies* **5**, 317–32.

Gershuny, J. 1987. Lifestyle, innovation and the future of work. *Royal Society of Arts Journal* **135**, 492–9.

Goad, C. E. 1987. *Goad register of out-of-centre shopping*. Old Hatfield, England: Goad.

Goldman, A. 1974a. Growth of large food stores in developing countries. *Journal of Retailing* **50**(2), 139–89.

Goldman, A. 1974b. Outreach of consumers and the modernization of urban food retailing in developing countries. *Journal of Marketing* **38**(4), 8–16.

Goldman, A. 1981. The transfer of retailing technology into the less developed countries: the supermarket case. *Journal of Retailing* **57**(2), 5–29.

Green Magazine 1991. High St checkout: sweet talking. *Green Magazine* (May), 65.

Guskind, R. & N. R. Pierce 1988. Faltering festivals. *National Journal* **20**, 2307–11.

Guy, C. M. 1980. *Retail location and retail planning in Britain*. Farnborough: Gower.

Guy, C. 1984. *Food and grocery shopping behaviour in Cardiff*. Working Paper 86, Department of Town Planning, University of Wales Institute of Science and Technology.

Guy, C. 1985. Some speculations on the retailing and planning implications of push-button shopping in Britain. *Environment and Planning* B **12**, 193–208.

Guy, C. 1987. Accessibility to multiple owned grocery stores in Cardiff: a description and evaluation of recent changes. *Planning Practice and Research* **2**, 9–15.

Guy, C. M. 1988a. Information technology and retailing: the implications for analysis and forecasting. In *Store choice, store location and market analysis*, N. Wrigley (ed.), 305–22. London: Routledge.

Guy, C. M. 1988b. Retail planning policy and the large grocery store development: a case study in South Wales. *Land Development Studies* **5**, 31–45.

Guy, C. M. 1992. Estimating shopping centre turnover: a review of survey methods. *Retail and Distribution Management* **20**, 18–23.

Guy, C. M. & J. D. Lord 1991. An international comparison of urban retail development. In *The attraction of retail locations, IGU Symposium August 1991*, G. Heinritz (ed.), vol. II, 120–33. München: Department of Geography, Technische

Universität München.
Guy, C. & N. Wrigley 1987. Walking trips to shops in British cities. *Town Planning Review* **58**, 63–79.

Hague, C. 1991. A review of planning theory in Britain. *Town Planning Review* **62**, 295–310.
Hakansson, H. (ed.) 1982. *International marketing and the purchasing of industrial goods*. Chichester, England: John Wiley.
Hall, P. 1985. The world and Europe. In *The future of urban form: the impact of new technology*, J. Brotchie, P. Newton, P. Hall, P. Nijkamp (eds), 17–30. London: Croom Helm.
Hall, P. 1988. *Cities of tomorrow: an intellectual history of urban planning and design in the twentieth century*. Oxford: Basil Blackwell.
Hallsworth, A. 1988. *Regional shopping centres: some lessons from Canada*. London: Transport and Environmental Studies.
Hallsworth, A. G. 1990a. The lure of the USA: some further reflections. *Environment and Planning* A **22**, 551–7.
Hallsworth, A. G. 1990b. More home thoughts from abroad. *Town and Country Planning* **59**, 51–3.
Hallsworth, A. G. 1991a. The Campeau takeovers – the arbitrage economy in action. *Environment and Planning* A **23**, 1217–24.
Hallsworth, A. G. 1991b. Regional shopping centres: case studies on recent policy decisions in Canada and the UK. *Service Industries Journal* **11**, 219–32.
Hamill, J. & J. Crosbie, 1990. British retail acquisitions in the US. *International Journal of Retail and Distribution Management* **18**(2), 15–20.
Harrison, H. 1990. Leisure facilities in shopping centres. *Shopping Centre Horizons* **20**, 13–20.
Hassington, J. 1985. Fear of crime in public environments. *Journal of Architecture and Planning Research* **2**, 289–300.
Harvey, D. 1982. *The limits to capital*. Oxford: Basil Blackwell.
Harvey, D. 1985. *The urbanization of capital*. Oxford: Basil Blackwell.
Heineberg, H. & A. Mayr 1988. Neue Standortgemeinschaften des großflächigen Einzelhandels im polyzentrisch strukturierten Ruhrgebiet. *Geographical Review* **40**, 28–38.
Henderson Crosthwaite 1992. *Three plus one*. Food retail report No 10. London: Henderson Crosthwaite Institutional Brokers.
Hillier Parker 1987. *Shopping centres of Great Britain. Master List*. London: Hillier Parker.
Hillier Parker 1991. *Shopping schemes in the pipeline*. London: Hillier Parker.
Hillman, M., I. Henderson, A. Whalley 1974. *Mobility and accessibility in the Outer Metropolitan Area, political and economic planning*. Report to the DOE, London: Policy Studies Institute.
Hollander, S. C. 1970. *Multinational retailing*. East Lansing: Michigan State University Press.
Home, R. K. 1991. Deregulating UK planning controls in the 1980s. *Cities* **8**, 292–300.
Home Office 1984. *The Shops Acts. Report of the Committee of Inquiry into proposals to amend the Shops Acts (Auld Committee)*. London: HMSO.
Home Office. 1988. *Home Office statement on safer cities programme*. London: Home Office.

Hopkins, J. S. P. 1991. West Edmonton Mall as a centre for social interaction. *Canadian Geographer* **35**, 268–79.
Howard, E. B. 1989. *Prospects for out-of-town retailing: the Metro experience*. Harlow, England: Longman.
Howard, E. B. (ed.) 1990. *Leisure and retailing*. Harlow, England: Longman.
Howard, E. B. & R. L. Davies 1988. *Change in the retail environment*. Harlow, England: Longman.
Howard, M. 1990. Out of town shopping: is the revolution over? *Royal Society for the Encouragement of Arts, Manufactures and Commerce, Journal* **5403**, 162–72.
Hurstfield, J. 1978. *The part-time trap*. London: Low Pay Unit.

Iacovone, L. 1990. *L'Associazionismo nella Distribuzione no Alimantaire*. Milan: CESCOM.
IFS (Institute of Fiscal Studies) 1984. Economic review conducted for the Home Office Committee of Inquiry into Proposals to Amend the Shops Acts. In *The Shops Acts: late night and Sunday opening*. Cmnd 9376, 101–99. London: HMSO.
Imrie, R. & P. Wells 1992. *Disability, access and the planning system: a study of South Wales*. SEAM technical report 1, Royal Holloway and Bedford New College, University of London.
IRESCO (Instituto de Reforma de las Estructuras Commerciales) 1984 *Informe sobre el Commercio Asociado en Espana*. Madrid: Instituto de Reforma de las Estructuras Commerciales, Ministerio de Economia y Hacienda.
Irvine, S. 1989. Consuming fashions? The limits of Green consumerism. *The Ecologist* **19**, 88–93.

Jackson, E. L. 1991. Shopping and leisure: implications of West Edmonton Mall for leisure and for leisure research. *Canadian Geographer* **35**, 280–7.
Jackson, E. L. & D. B Johnson 1991. Geographic implications of mega-malls with special reference to West Edmonton Mall. *Canadian Geographer* **35**, 226–32.
Jackson, G. I. 1973. Planning the move to the Continent: questions that must be asked. *Retail and Distribution Management* **1**(6), 14–16.
Jacobs, B. D. 1992. *Fractured cities*. London: Routledge.
Jacobs, J. 1961. *The death and life of great American cities*. New York: Vintage Books.
Jansen, A.C.M. 1989. Funshopping as a geographical notion, or: the attraction of the inner city of Amsterdam as a shopping area. *Tijdschrift voor Economische en Sociale Geografie* **80**, 171–83.
Jarvis, V. J. & S. J. Prais 1989. *Two nations of shopkeepers: training for retailing in France and Britain*. National Institute Economic Review 128. London: National Institute for Economic and Social Research.
Jefferys, J. 1973. Multinational retailing: are the food chains different. *CIES Quarterly Review* **8**(3).
Johnson, D. B. 1991. Structural features of West Edmonton Mall. *Canadian Geographer* **35**, 249–61.
Johnson, G. 1987a. Introduction. In *Business strategy and retailing*, G. Johnson (ed.), 1–5. Chichester, England: John Wiley.
Johnson, G. 1987b. Strategies and strategic positioning in retailing. In *Business strategy and retailing*, G. Johnson (ed.), 81–7. Chichester, England: John Wiley.
Johnson, S. 1990. The leisure market: consumer choice and consumer activity. Section 3, in *Leisure and retailing*, Howard, E. (ed.). Harlow, England: Longman.
Jones, K. & J. Simmons 1990. *The retail environment*. London: Routledge.

Jones, P. 1981. Retail innovation and diffusion: the spread of Asda stores. *Area* **13**, 197–201.

Jones, P. 1989. The High Street fights back. *Town and Country Planning* **58**, 43–5.

Jones, P. 1991. Regional shopping centres: the planning issues. *Service Industries Journal* **11**, 171–8.

Kacker, M. P. 1985. *Transatlantic trends in retailing*. Westport, Connecticut: Greenwood.

Kacker, M. P. 1988. International flow of retailing know-how: bridging the technology gap in distribution. *Journal of Retailing* **64**(1), 41–67.

Kacker, M. P. 1990. The lure of US retailing to the foreign acquirer. *Mergers & Acquisitions* **25**(1), 63–8.

Kay, J. A., C. N. Morris, S. M. Jaffer, S. A. Meadowcroft 1984. *Effects of Sunday trading: the regulation of trading hours*. London: The Institute for Fiscal Studies.

Kaye, T. 1988. Retail and leisure – myth and reality. *Estates Gazette* **8829**, 53–62.

Kern, C. R., S. R. Lerman, R. J. Parcells, R. A. Wolfe 1984. *Impact of transportation policy on the spatial distribution of retail activity*. Washington, DC: US Department of Transportation.

Kleinwort Benson 1990. *Aldi: the eagle has landed*. London: Kleinwort Benson Securities.

Knee, D. 1966. Trends towards international operations among large-scale retailing enterprises. *Rivista Italiana di Amministrazione* **2**, 107–11.

Knee, D. 1968. European retail trade associations come of age. *Journal of Retailing* **44**(1), 13–28.

Knee, D. 1988. *City centre retailing in continental Europe*. Harlow, England: Longman.

Laulajainen, R. 1987. *Spatial strategies in retailing*. Dordecht: Reidel.

Laulajainen, R. 1991a. International expansion of an apparel retailer: Hennes and Mauritz of Sweden. *Zeitschrift für Wirtschaftsgeographie* **35**(1), 1–15.

Laulajainen, R. 1991b. Two retailers go global: the geographical dimension. *International Review of Retail Distribution and Consumer Research* **1**(5), 607–26.

Laulajainen, R. 1992. Louis Vuitton Malletier: a truly global retailer. *Japan Association of Economic Geographers, Annals* **38**(2), 55–70.

Lawless, P. 1989. *Britain's inner cities: problems and policies*. London: Paul Chapman.

Lawless, P. & C. Raban (eds) 1986. *The contemporary British city*. London: Harper & Row.

LDDC (London Docklands Development Corporation) 1988. *Annual review 1987–88*. London: LDDC.

Lee Donaldson Associates 1979. *Caerphilly hypermarket study, year five*. Donaldsons Research Report 6. London: Lee Donaldson Associates.

Leicester City Council and Leicestershire Constabulary 1989. *Crime prevention by design*. Leicester.

Leontiades, J. C. 1985. *Multinational corporate strategy*. Lexington, Mass.: Lexington Books.

Lichfield, D. 1990. From combination to integration. Section 5 in *Leisure and retailing*, Howard, E. (ed.). Harlow, England: Longman.

Litwak, D. 1987. The nation's top grocery corporations. *Supermarket Business* **42**(3), 92–6.

London Borough of Brent 1986. *Tesco, Neasden: a study of retail impact*. London: London Borough of Brent.

Lord, D., W. Moran, A. Parker, L. Sparks 1988. Retailing on three continents: the discount food store operations of Albert Gubay. *International Journal of Retailing* **3**(3), 1–54.

Lord, J. D. 1988. *Retail decentralization and CBD decline in American cities*. Working Paper 8802, Institute for Retail Studies, University of Stirling.

Lord, J. D. & C. M. Guy 1991. Comparative retail structure of British and American cities: Cardiff (UK) and Charlotte (USA). *International Review of Retail, Distribution and Consumer Research* **1**, 391–436.

Lowe, M.S. & L. Crewe 1991. Lollipop jobs for pin money? Retail employment explored. *Area* **23**, 344–7.

McGoldrick, P. M. 1990. *Retail marketing*. Maidenhead, England: McGraw-Hill.

McKay, G. & A. Corke 1986. *The Body Shop: franchising a philosophy*. London: Pan.

McKay, J. & K. Fletcher 1988. Consumers' attitudes towards teleshopping. *Quarterly Review of Marketing (UK)* **13**(3), 1–7.

McKinnon, A. C. 1989. *Physical distribution systems*. London: Routledge.

McLauglin, B. 1986. The rhetoric and the reality of rural deprivation. *Journal of Rural Studies* **2**, 291–307.

Magowan, P. 1989. The case for LBO's: the Safeway experience. *California Management Review* **32**, 9–18.

Manchester University 1988. *UK home shopping 1988*. Manchester: Centre for Business Research, Manchester Business School.

Marion, B. W. (ed) 1986. *The organization and performance of the US food industry*. Lexington, Mass.: Lexington Books.

Marks & Spencer plc 1991. *Annual report and financial statements*. London: Marks & Spencer.

Martenson, R. 1981. *Innovation in multinational retailing*. Gothenburg: University of Gothenburg.

Martenson, R. 1987. Culture bound industries? A European case study. *International Marketing Review* **4**(3), 7–17.

Martenson, R. 1988. Cross-cultural similarities and differences in multinational retailing. In *Transnational retailing*, E. Kaynak (ed.). Berlin: Walter de Gruyter.

Marti, J. & A. Zeilinger 1982. Videotext in the High Street. *Retail and Distribution Management* **10**(5), 21–6.

Martin, J. & C. Roberts 1984. *Women and employment: a lifetime perspective*. The report of the 1980 DE/OPCS women and employment survey. London: HMSO.

Martinez, J. M. 1991. La Consolidación. *Distribuçión Actualidad* **176**, 35–7.

Mathias, P. 1967. *Retailing revolution*. London: Longman.

Mazur, A. 1991. *Downtown development – Chicago: 1989–1992*. Chicago: City of Chicago.

Meadows, D. H., D. L. Meadows, J. Randers, W. W. Behrens III 1972. *The limits to growth: a report for the Club of Rome's project on the predicament of mankind*. London: Earth Island.

Meier, R. 1985. High tech and urban settlement. In *The future of urban form: the impact of new technology*, J. Brotchie, P. Newton, P. Hall, P. Nijkamp (eds), 70–75. London: Croom Helm.

Merenne-Schoumaker, B. 1983. Libre-service et centres commerciaux en Europe. Evolution récente. *Bulletin de la Société Géographique de Liège* **19**, 63–76.

Merenne-Schoumaker, B. 1991. *From hypermarkets to shopping centres. The peripheral poles coming to light*. Paper presented at International Symposium of the IGU Commission, "Geography of commercial activities", München.

Metton, A. 1983. Centres commerciaux périphériques: regards sur l'expérience parisienne. *Revue de Géographie des Pyrénées et du Sud-Ouest* **5**, 89–110.

Milder, N. D. 1987. Crime and downtown revitalization. *Urban Land* (September), 16–25.

Min, H. & W. P. Galle 1991. International purchasing strategies of multinational US firms. *International Journal of Materials and Purchasing Management*, **5**(4), 9–18.

Ministry of the Environment 1991. *Developing the city centres as shopping areas*. Helsinki: Government Publishing Office.

Mishan, E. J. 1990. Economic and political obstacles to environmental sanity. *National Westminster Bank Quarterly Review* (May), 25–42.

MMC (Monopolies and Mergers Commission) 1981. *Discounts to retailers* HC311. London: HMSO.

Moir, C. 1990. Competition in the UK grocery trades. In *Competition and markets: essays in honour of Margaret Hall*, C. Moir & J. Dawson (eds), 91–118. London: Macmillan.

Monczka, R. M. & R. J. Trent 1991. Evolving sourcing strategies for the 1990s. *International Journal of Logistics and Distribution Management* **21**(5), 4–12.

Moore, V. 1990. *A practical approach to planning law*. London: Blackstone Press.

Morgan, A. D. 1988. *British imports of consumer goods*. Cambridge: Cambridge University Press.

Morganosky, M. & M. Lazarde, 1987. Foreign made apparel: influences on consumer's perceptions of brand and store quality. *International Journal of Advertising* **6**, 339–46.

Morphet, J. 1991. Town centre management - old problems, new solutions. *Town and Country Planning* **60**, 202–3.

Morrill, R. L. 1987. The structure of shopping in a metropolis. *Urban Geography* **8**, 97–128.

Murphy, R. E. & J. E. Vance Jr 1954. Delimiting the CBD. *Economic Geography* **30**, 189–222.

Nathan, R. & P. Dommel 1977. Understanding the urban predicament. *The Brookings Bulletin* **14**, 9–13.

National Association of Local Government Women's Committees 1991. *Responding with authority*. London: NALGWC.

National Audit Office 1990. *Regenerating the inner cities*. London: HMSO.

NCC (National Consumer Council) 1991. *Trade and competition policy: the consumer interest. International trade and the consumer*. Working Paper 5. London: National Consumer Council.

NEDO (National Economic Development Office) 1988a. *The future of the High Street*. London: NEDO.

NEDO (National Economic Development Office) 1988b. *Part-time working in the distributive trades, training practices and career opportunities*, Volumes 1 & 2. London: NEDO.

Nelson, R. L. 1958. *The selection of retail locations*. New York: Dodge.

Newman, O. 1972. *Defensible space: people and design in the violent city*. London: Architectural Press.

Newby, P. 1991. *Leisure and shopping in the modern mall: summary of survey*. London:

Middlesex Polytechnic School of Geography and Planning.
Nisse, J. 1991. BSkyB may need extra £400 million. *The Independent* (25 May), 21.
Norkett, P. 1985. Stack 'em high, sell 'em fast: the key to supermarket success. *Accountancy* **96**, 74–9.
Nottingham City Council 1990. *Nottingham safer cities project*: Steering Group report on safety in the city centre. Nottingham: Nottingham City Council.

O'Brien, L. G. & F. W. Harris 1991. *Retailing: shopping, society, space*. London: David Fulton.
OFT (Office of Fair Trading) 1985. *Competition and retailing*. London: Office of Fair Trading.
Ogg, A. 1990. *Proceedings: third retail and leisure conference*. Manchester: Leisure and Retail Communications.
OPCS (Office of Population Censuses and Surveys) 1987. *General household survey 1986*. London: HMSO.
OPCS (Office of Population Censuses and Surveys) 1991. *General household survey 1989*. London: HMSO.
O'Riordan, T. 1984. *Environmentalism*. London: Pion.

Panorama 1991. *Panorama 92: Points de vente. La recensement de la Grande Distribution au 01/09/91*. Paris: Liaisons & Convergence.
Paumier, C. B. 1982. Design elements for a downtown market place. *Journal of Housing* (May/June), 76–9.
Paumier, C. B. 1988. *Designing the successful downtown*. Washington, DC: The Urban Land Institute.
Peattie, K. 1992. *Green marketing*. London: Pitman.
Peet, R. 1991. *Global capitalism*. London: Routledge.
Pellegrini, L. 1992. The internationalization of retailing and 1992 Europe. *Journal of Marketing Channels* 1(2), 3–27.
Penny, N. J. 1984. A geographical study of intra-urban shopping behaviour in greater Swansea. PhD thesis, University of Wales, Swansea.
Percival, A. 1991. The developer in Europe. Paper presented at "Europe: the emergence of the world's largest retail market". London: IBC/Debenham/Jean Thouard Zadelhoff.
Perkins, S. 1992. Retail revolution in the Black Country. Dissertation, Department of Geography, University College of Swansea.
Phillips, D. & A. Williams 1984. *Rural Britain: a social geography*. Oxford: Basil Blackwell.
Pickup, L. 1984. *Women's gender role and its influence on travel behaviour*. Working Paper 10(1), University of Reading.
Pickup, L. 1988. Hard to get around: a study of women's travel mobility. In *Women in cities, gender and the urban environment*, J. Little, L. Peak, P. Richardson (eds), 88–116. London: Macmillan.
Pond, C. 1977. *Trouble in store*. London: Low Pay Unit.
Ponting, C. 1991. *A Green history of the world*. Harmondsworth: Penguin.
Poole, R. & K. Donovan 1991. *Safer shopping. The identification of opportunities for crime and disorder in covered shopping centres*. Birmingham: West Midlands Police & Home Office Police Requirements Support Unit.
Porritt, J. 1986. *Seeing Green*. Oxford: Basil Blackwell.
Porritt, J. (ed.) 1987. *Friends of the Earth handbook*. London: Optima.

Porritt, J. 1990. *Where on Earth are we going?* London: BBC Books.

Porritt, J. & D. Winner 1988. *The coming of the Greens*. London: Fontana Collins.

PPS (Project for Public Spaces) 1984. *Managing downtown public spaces*. Chicago: APA Publications.

Price, J. & B. Yandle 1987. Labor markets and Sunday closing laws. *Journal of Labor Research* **7**, 407–14.

Ralphs, G. 1989. Development control and the mediated negotiation: A case study of Cardiff. MSc dissertation, University of Wales, College of Cardiff.

Rees, G. 1969. *St Michael: a history of Marks & Spencer*. London: Weidenfeld & Nicolson.

Rees, J. C. M. 1986. Retail growth in the Swansea Enterprise Zone: a study of the impact of an out-of-town shopping centre. Dissertation, Department of Town Planning, Bristol Polytechnic.

Rees, J. 1987. Perspectives on retail planning issues. *Planning Practice and Research* **2**, 3–8.

Reid, V. 1991. Out of town offices hurt the town centre? *Property Research Summaries* 12–13. London: Hillier Parker Research.

Retail Business 1988. Teleshopping update. *Retail Business* **370**, 4–9.

Retail and Distribution Management 1986. Bradford's new teleshopping service. *Retail and Distribution Management* **14**(1), 35–6.

Retail and Distribution Management 1988. Teleshopping scheme started by Asda. *Retail and Distribution Management* **16**(3), 42.

Reynolds, J. 1983. Retail employment change: scarce evidence in an environment of change. *Service Industries Journal* **3**, 334–62.

Reynolds, J. 1990. Cycles in property costs. Section 4 in *Retailing in the 1990s: threats to profitability*, 53–70. Harlow, England: Longman.

Reynolds, J. & R. Schiller 1992. A new classification of shopping centres in Great Britain using multiple branch numbers. *Journal of Property Research* **9**, 89–110.

Richardson, H. W. 1978. *Urban economics*. Hinsdale, Illinois: Dryden.

Roberts, J. & R. Frampton 1991. Conference report Nice 1991. *Shopping Centre Horizons* (spring), 3–13.

Roberts, J. & N. James 1990. Some transport is more equal than others. *The Planner* **76**(12), 22–5.

Roberts, M. 1990. Gender and housing: the impact of design. *Built Environment* **16**, 262–75.

Robertson, K. 1983. Downtown retail activity in large American cities 1954–1979. *Geographical Review* **73**, 314–23.

Robertson, K. 1990. The status of the pedestrian mall in American downtowns. *Urban Affairs Quarterly* **26**, 250–73.

Robertson, P. 1991. Assistants pressurized into Sabbath working. *The Journal* **16** (December).

Robinson, O. 1990. Employment policies in the service sector: training in retail distribution. *Service Industries Journal* **10**, 284–305.

Robinson, O. & Wallace, J. 1976. *Pay and employment in retailing*. Farnborough: Saxon House.

Robson, B. 1988. *Those inner cities: reconciling the social and economic aims of urban policy*. Oxford: Oxford University Press.

Robinson, T. M. & C. M. Clarke-Hill 1990. Directional growth by European retailers. *International Journal of Retail and Distribution Management* **18**(5), 3–14.

REFERENCES

Roddick, A. 1991. *Body and soul*. London: Ebury Press.

Rojek, C. 1990. Baudrillard and leisure. *Leisure Studies* **9**, 7–20.

Rossi, I. 1991. Il commercio riscopre la galleria. *Largo Consumo* **3**, 141–7.

Rowley, G. 1984. Local government or central government agency? – The British case. *Political Geography Quarterly* **4**, 259–64.

Rowley, G. 1985. Superstores and hypermarkets: databases on out-of-centre developments. *Institute of British Geographers, Transactions* **10**, 380–2.

Rowley, G. 1986. *Let's talk shop: relocational trends in British retailing*. Old Hatfield, England: Goad.

Rowley, G. 1989. *Impact assessment of an out-of-centre shopping centre: Meadowhall, Sheffield*. Department of Town and Regional Planning Occasional Papers 86, University of Sheffield.

Rowley, G. 1990. City centre retail structures in Britain: continuing development or fundamental change? In *Papers on retail development and consumer behaviour: proceedings of the Oxford symposium*, E. Howard (ed.), 37–51. Oxford: Templeton College.

Rowley, G. 1992. *Urban spatial growth and metropolitan areas: a consideration of the dynamics, components and developing problems of British cities*. Department of Town and Regional Planning Occasional Papers 110, University of Sheffield.

RTPI (The Royal Town Planning Institute) 1988. *Planning for shopping in the 21st century: the report of the retail planning working party*. London: RTPI.

Sainsbury, J. plc 1991. *Some facts about Sainsbury's*. London: J. Sainsbury plc.

Sainsbury, J. plc 1992. *Living today, the environment*. London: J. Sainsbury plc.

Salmon, W. J. & A. Tordjman 1989. The internationalization of retailing. *International Journal of Retailing* **4**(2), 3–16.

Savitt, R. 1990. Retail change and economic development. In *Retailing environments in developing countries*, A. M. Findlay, R. Paddison & J. A. Dawson (eds.), 16–29. London: Routledge.

Sawicki, D. S. 1989. The festival marketplace as public policy: guidelines for future policy decisions. *American Planning Association, Journal* **55**, 347–61.

Sayer, A. 1985. The geography of industry: multinationals. In *Changing world: geographical Perspectives*, D205 Changing Britain, Unit 5, Block 2, 1–25. Milton Keynes: Open University.

Scarth, W. & A. Ashcroft 1992. Wolverhampton draws quarter plan. *Planning* **965**, 8.

Schiller, R. 1985. Land use controls on UK shopping centres. In *Shopping centre development: policies and prospects*. J. A. Dawson & J. D. Lord (eds), 40–56. Beckenham: Croom Helm.

Schiller, R. 1986. Retail decentralization – the coming of the third wave. *The Planner* **72**, 13–15.

Schwartz, G. G. 1984. *Where is Main Street, USA?* Westport, Conn.: ENO Foundation.

Segal-Horn, S. & J. McGee 1989. Strategies to cope with retailer buying power. In *Retail and marketing channels: economic and marketing perspectives on producer-distributor relationships*, L. Pellegrini & S. Reddy (eds), 28–48. London: Routledge.

Sennett, R. 1991. *The conscience of the eye: the design and social life of cities*. London: Faber & Faber.

Shaw, S. A., J. A. Dawson, L. M. A. Blair 1992. Imported foods in a British supermarket chain: buyer decisions in Safeway. *International Review of Retail,*

Distribution and Consumer Research **2**(1), 35–57.

Shepherd, P. McL. & G. Rowley 1978. The association of retail functions within the city centre. *Tijdschrift voor Economische en Sociale Geographie* **69**, 233–7.

Shiret, T. 1992. *How much hot air do you like in your accounts? Food retailer property valuations and their effect on the P & L.* London: Credit Lyonnais Laing.

Siegle, N. & C. R. Handy 1981. Foreign ownership in food retailing. *National Food Review*, **3**(2), 14–16.

Silbertson, A. 1990. Textile markets and the Multi-Fibre Arrangements. In *Competition and markets*, C. Moir & J. A. Dawson (eds), 63–76. Basingstoke: Macmillan.

Smith New Court 1991. *The food retail sector: returning to the buy tack*. London: Smith New Court Securities.

South Glamorgan County Council 1990. *South Glamorgan retail floorspace survey 1990*. Cardiff: South Glamorgan County Council.

Sparks, L. 1983. Employment characteristics of superstore retailing. *Service Industries Journal* **3**(1), 63–78.

Sparks, L. 1986. The changing structure of distribution in retail companies: an example from the grocery trade. *Institute of British Geographers, Transactions* **11**, 147–54.

Sparks, L. 1987. Employment in retailing: trends and issues. In *Business strategy and retailing*, G. Johnson (ed.), 239–55. Chichester, England: John Wiley.

Sparks, L. 1990. Spatial-structural relationships in retail corporate growth: a case study of Kwik Save Group plc. *The Service Industries Journal* **10**, 25–84.

Sparks, L. 1992. Restructuring retail employment. *International Journal of Retail and Distribution Management* **20**(3), 12–19.

Staniland Hall Associates 1992. *UK retailing 1992–1996. Sector forecasts and management issues*. Slough/Oxford: Staniland Hall Associates / Oxford Institute of Retail Management.

Sutcliffe, P. 1988. New ingredients in the retail brew. *Estates Gazette* **8829**, 63.

Tager, U. C. & G. Weitzel 1991. *Purchasing organizations*. DG XXIII, Series Studies, Commerce and Distribution, 19. Brussels: Commission of the European Communities.

Taylor, A. 1984. The planning implications of new technology in retailing and distribution. *Town Planning Review* **55**, 161–76.

Taylor, P. J. 1991. Understanding global inequalities: a world-systems approach. *Geography* **77**, 10–21.

Telling, A. 1990. *Planning law and procedure*, 8th edn. London: Butterworths.

TEST (Transport & Environmental Studies) 1989. *Trouble in store? Retail locational policy in Britain and Germany*. London: TEST.

Thil, E. 1966. *Les inventeurs du commerce moderne*. Paris: Arthaud.

Thomas, C. J. 1974. The effects of social class and car ownership on intra-urban shopping behaviour in Greater Swansea. *Cambria, a Welsh Geographical Review* **1**, 98–126.

Thomas, C. J. 1977. *Leo's superstore, Pyle, Mid-Glamorgan*. Retail Outlets Research Unit, Research Report 22, Manchester Business School.

Thomas, C. J. 1989. Retail change in greater Swansea: evolution or revolution? *Geography* **71**, 201–13.

Thomas, C. J. & R. D. F. Bromley 1987. The growth and functioning of an unplanned retail park: the Swansea Enterprise Zone. *Regional Studies* **21**, 287–300.

Thomas, H. 1978. Costing the value of ecological disasters. *The Guardian*, 19 July. Cited in *Modern Western society*, P. Dicken & P. E. Lloyd 1981. London: Harper & Row.

Thompson, R. 1990. Planning for the have-nots. *The Planner* **76**(12), 10–12.

Thornley, A. 1990. *Urban planning under Thatcherism*. London: Routledge.

Thorpe, D. 1991. The development of British superstore retailing – further comments on Davies and Sparks. *Institute of British Geographers, Transactions* **16**, 354–67.

Thorpe, D. 1992. *The changing geography of British shopping and retailing 1945–1991: an overview and an agenda*. Paper presented at the Institute of British Geographers conference, Swansea.

Thorpe, D. et al. 1972. *The Hampshire Centre, Bournemouth*. Retail Outlets Research Unit, Research Report 3, Manchester Business School.

Tilley, L. 1989. Presto! France's Minitel success. *The Times* (21 December), 33.

Tonion, S. 1991. A travers les dédales italiens. *Sites Commerciaux*, March.

Tordjman, A., J. A. Dawson, B. Gazan 1993. *The European retailing industry*. DG XXIII. Brussels: Commission of the European Communities.

Tordjman, A. & J. Dionisio 1991. *Internationalization strategies of retail business*. DG XXIII, Series Studies, Commerce and Distribution, 15. Brussels: Commission of the European Communities.

Townsend, A. R. 1986. Spatial aspects of the growth of part-time employment in Britain. *Regional Studies* **20**, 313–30.

Traill, B. (ed.) 1989. *Prospects for the European food system*. London: Elsevier.

Treadgold, A. 1990a. *The costs of retailing in continental Europe*. Harlow, England: Longman.

Treadgold, A. 1990b. The developing internationalization of retailing. *International Journal of Retail and Distribution Management* **18**(2), 4–11.

Treadgold, A. 1991. The emerging internationalization of retailing: present status and future challenges. *Irish Marketing Review* **5**(2), 11–27.

Treadgold, A. & R. L. Davies 1988. *The internationalization of retailing*. Harlow, England: Longman.

Toronto City Council 1988. *The safe city – municipal strategies for preventing public violence against women*. Toronto: Toronto City Council.

Trench, S., T. Oc, S. Tiesdell 1992. Safer cities for women – perceived risks and planning measures. *Town Planning Review* **63**, 279–96.

UCD (University College, Dublin) 1990. *Dublin shopping centres: statistical digest II*. Dublin: UCD, The Centre for Retail Studies.

ULI (Urban Land Institute) 1983. *Revitalizing downtown retailing – trends and opportunities*. Washington, DC: ULI.

Usher, D. 1980. *The measurement of economic growth*. Oxford: Basil Blackwell.

UN (United Nations) 1985. *Transnational trading corporations in selected Asian and Pacific countries*. ESCAP, UNCTC, Series B, 6. New York: United Nations.

Valentine, G. 1990. Women's fear and the design of public space. *Built Environment* **16**, 288–303.

Wade, B. 1985. New directions in retailing. In *The future of planning: planning for retailing*, Greater London Council (ed.), 11–15. London: Greater London Council.

Waldman, C. 1978. *Strategies of international mass retailers*. New York: Praeger.

Wang, C. & C. Lamb 1983. The impact of various environmental forces upon consumers' willingness to buy foreign products. *Journal of the Academy of Marketing Science* **11**(2), 71–84.

Ward, D. 1966. The industrial revolution and the emergence of Boston's central business district. *Economic Geography* **42**, 152–7.

Warnes, A. M. 1989. Social problems of elderly people in cities. In *Social problems and the city*, D. T. Herbert & D. M. Smith (ed.), 197–212. Oxford: Oxford University Press.

Warren, G. & D. Taylor 1991. The big shop – carless in Camden. *Town and Country Planning* **60**, 206–7.

Watson, G. 1992. Hours in work in Great Britain and Europe. *Employment Gazette* **100**, 539–57.

Welch, L. S. & R. Luostarinen 1988. Internationalization: evolution of a concept. *Journal of General Management* **14**(2), 35–55.

Westlake, T. 1990. Electronic home shopping: do planners need to know? *Planning Practice and Research* **5**(1), 6–10.

Westlake, T. 1992. The planning implications of interactive viewdata systems: a study of electronic home shopping. PhD thesis, University of Wales, College of Cardiff.

Westlake, T. & K. Dalgleish 1990. *Disadvantaged consumers: do planners care?* Working Paper 44, School of Planning, Birmingham Polytechnic.

Which? 1991. Teleshopping. *Which?* (March), 148–9.

Whitehead, M. 1992. Internationalization of retailing: developing new perspectives. *European Journal of Marketing* **26**(8/9), 74–9.

Wilkins, M. 1974. *The maturing of multinational enterprise*. Cambridge, Mass.: Harvard University Press.

Williams, D. E. 1992. Retailer internationalization: an empirical inquiry. *European Journal of Marketing* **26**(8/9), 8–24.

Williams J. J. 1991. Meadowhall: its impact on Sheffield city centre and Rotherham. *International Journal of Retail and Distribution Management* **19**(1), 29–37.

Wilson, E. 1991. *The sphinx in the city: the control of disorder and women*. London: Virago.

Wittstock, M. 1990. Cable subscribers can buy food by television keypad. *The Times* (11 September), 3.

Wrigley, N. 1984. Geographical evidence to the Committee of Inquiry on the Shops Act. *Area* **16**, 223–43.

Wrigley, N. 1987. The concentration of capital in UK grocery retailing. *Environment and Planning* A **19**, 1283–8.

Wrigley, N. 1988. Retail restructuring and retail analysis. In *Store choice, store location and market analysis*, N. Wrigley (ed.), 3–34. London: Routledge.

Wrigley, N. 1989. The lure of the USA: further reflections on the internationalization of British grocery retailing capital. *Environment and Planning* A **21**, 283–8.

Wrigley, N. 1991. Is the "golden age" of British grocery retailing at a watershed? *Environment and Planning* A **23**, 1537–44.

Wrigley, N. 1992a. Antitrust regulation and the restructuring of grocery retailing in Britain and the USA. *Environment and Planning* A **24**, 727–49.

Wrigley, N. 1992b. Sunk capital, the property crisis and the restructuring of British food retailing. *Environment and Planning* A **24**, 1521–7.

Yearley, S. 1991. *The Green case: a sociology of environmental issues, arguments and poli-*

tics. London: HarperCollinsAcademic.
Yeates, M. & B. Garner 1980. *The North American city*. New York: Harper & Row.
Yoshino, M. Y. 1966. International opportunities for American retailers. *Journal of Retailing* (fall), 1–10.

Zentes, J. & W. Schwarz-Zanetti, 1988. Planning for retail change in West Germany. *Built Environment* **14**, 38–46.

INDEX

A6 centre 86
access 3, 7, 15, 18, 29, 52, 58, 65, 67, 88, 103, 112, 114, 124, 126, 127, 133, 143, 145, 150, 162, 164, 165, 168, 174, 176–82, 184–90, 246, 253–5
accessibility 10, 112, 114, 148, 151, 154, 157, 174, 178, 180, 181, 252, 253
adjacency 112
affluence 3, 6, 209, 240, 247
agglomeration economies 112, 113
ambience 215, 221
anchor stores 72, 79–82, 86, 101–103, 121, 132, 142–4, 218
antiques 82, 222
apparel (*see also* clothing) 91, 114–16
arbitrage 58

Brent Cross 141, 217
Broadmead, Bristol 122, 123, 147
bulk-purchasing 127, 175
business parks 116, 118
buyer decisions 9, 17, 18
buying groups 20, 21, 23

capital switching 249
capitalism 229, 231, 233–5, 238, 246, 257
Capitol Exchange 100, 101, 105, 108
car
  boot sales 224
  ownership 3, 6, 10, 70, 134, 147, 174, 175, 177, 247, 250
  parks 12, 92, 105, 115, 139, 146, 164–6, 253, 254
Cardiff 8, 10, 89–91, 90–92, 94, 95, 94–102, 104, 107, 108, 146, 251
carers 13, 198, 199
carpets 7, 96, 136–8
central business district (CBD) 6, 8, 11, 98, 110–16, 118, 119, 121, 123–5, 143, 147, 148, 154, 180, 252
Centrepoint, Bradford 184
centres intercommunaux 79–81
charity shops 118
Charlotte, North Carolina 10, 89–91, 90–95, 97–9, 102, 105–108, 251
city centres 3, 10, 11, 77, 88–92, 93, 94, 95, 96, 97, 98, 99, 100, 102–105, 107, 108, 110, 111, 112, 116, 118, 121, 122, 124, 126, 129, 131, 132, 133, 135–41, 142, 143, 145–50, 153, 154–8, 160–62, 164–9, 174, 251–5, 258
clothing 7, 17, 19, 20, 79, 91, 95, 100, 118, 136–8, 140, 142, 233, 235
combination centre 219
commerciaux régionaux 81, 82
compaction 143–5, 147, 252
concentric zonal model 112
consumer
  behaviour 125, 126 (*see also* shopping behaviour)
  deprivation 176, 179, 190
  frustration 208, 216, 257
consumerism 229, 230, 240
consumers 4, 6, 12, 13, 15, 20, 27, 36, 53, 72, 79, 113, 115, 133, 138, 141, 147, 172, 173, 174–7, 179–82, 186–90, 205, 206, 216, 227, 229, 230, 233, 234, 236, 239, 240–42, 244–7, 254–8
  neglected 105, 173, 181, 186, 189
convenience
  goods 6, 144, 172, 190
  shopping 88, 173, 177, 190, 216
counter-urbanization 3, 6, 111, 244
Cribbs Causeway, Bristol 122, 123
Crystal Peaks, Sheffield 220
Culverhouse Cross, Cardiff 8

de-skilling 196, 197
decentralization 3, 8, 70, 88, 94, 108, 125, 134, 138, 140, 146, 147, 149–51, 179, 250–52, 253, 258
deindustrialization 4
department stores 8, 15, 20, 30, 79, 81, 88, 90, 96–8, 102, 103, 105, 108, 118, 144,

149, 196
deprivation 175, 176, 179, 190
deprived consumers 173, 179, 186, 189
design 12, 18, 20, 38, 77, 103, 105, 156, 165, 167, 169, 172, 180, 181, 214, 219, 220, 222, 224–8, 235, 254, 255, 257
Direct Foreign Investment (DFI) 27–9
disabled 12, 172–4, 178, 180–82, 184, 189–91, 193, 256
disadvantage 4, 173, 176, 256
disadvantaged consumers 12, 172–5, 179–82, 187, 189, 190, 247, 255, 256
discount food warehouses 65
discounters 9, 53, 54, 61, 62, 65–7, 116, 118, 190
disinvestment 42, 98
distribution 2, 3, 5, 10, 18, 24, 26, 36, 45, 59, 60, 73, 82, 91, 95, 98, 164, 172, 179, 182, 184, 185, 236, 239, 242, 245, 247, 258
district centres 6, 127, 77, 131, 132, 136–40, 144, 145, 150
DIY 5, 7, 18, 29, 30, 79, 96, 115, 117, 134–9, 196, 204, 210, 252
downtown 98, 103, 107, 111, 113, 116, 124, 147–9, 154, 157, 159, 161

elderly people 3, 12, 111, 132, 155, 158, 167, 172–5, 178, 179, 181, 183, 184, 188, 189, 191, 199, 206
Eldon Square, Newcastle upon Tyne 81, 121, 143, 145
electrical goods 7, 115, 138, 252
electronic
  data interchange (EDI) 20
  funds transfer (EFTPOS) 47
  home shopping (EHS) 5, 12, 172, 174, 179, 182, 183, 185–91, 256
entertainment provision 226
environmental
  impact 126, 237, 239, 240, 248
  policies 10, 236
electronic point-of-sale (EPOS) 5, 47
ethnic
  dimension 147
  minorities 12, 111, 172, 173, 175, 192
European Community (EC) 10, 19, 34, 35, 39, 52, 63, 124, 195, 205, 206
expertise 9, 15, 16, 28, 36, 38

fachmarkte 79
Factory Acts 209
factory outlet 72, 81, 82, 106
fashion 35, 60, 79, 119, 217, 221, 235
fear of crime 154, 155, 161, 254
festival shopping 10, 89, 108, 223, 257
footwear 7, 91, 95, 118, 136–8, 140, 196
Fosse Park, Leicester 81, 118
free-market ideology 5
functional shopping 212, 213, 217
functionalism 153
functionalist planning 153, 165
furniture 7, 38, 79, 91, 96, 114, 118, 123, 135–40, 211, 220, 244

Galleria, Hatfield 218
garden centres 115, 134, 135, 214, 215, 252
General Agreement on Tariffs and Trade GATT 19
golden age (of retailing) 6, 41, 43, 48, 51, 54, 55, 70, 102
Green
  awareness 13, 229, 230, 236, 237, 243, 244
  capitalism 229
  consumers 13, 229, 231–3, 239, 240, 243–5
  movement 13
  retailing 13, 230, 232
grocery 4, 5, 7, 9, 15, 21, 23, 41–4, 47–55, 57, 58, 60–63, 65–7, 96, 114, 126–34, 136, 150, 185, 189, 194, 196, 248, 249, 252

Haydock Park, Lancashire 141
hierarchy 6, 8, 80, 121
hypermarkets 7, 17, 28, 30, 32, 33, 79, 81, 84, 86, 126, 127

impacts (retail) 4, 11, 19, 20, 40, 61, 65, 66, 76, 126–8, 131–6, 138, 140–43, 147, 149, 150–52, 168, 175, 182, 186, 188, 203–205, 216, 237, 239, 240, 247, 248, 250, 252, 253, 255
industrial capitalism 231
information technology 5, 26, 29, 47, 182, 189, 249, 258
inner cities 11, 12, 41, 42, 110, 111, 123–5, 129, 130, 141, 147, 150, 155, 173, 175,

182, 252
intermediate centres 80, 81
international
operations 9, 27, 30–32, 34, 35, 39, 40, 248
sourcing 9, 15–17, 19–21, 25, 26, 40, 248
internationalization 4, 5, 9, 10, 15, 16, 18–20, 27–30, 32, 34, 36–8, 40–43, 49, 51, 54, 60, 66, 67, 68, 71, 248, 249
inward movement of retail capital 42, 67

jewellery shopping/shops 115, 214

labour
organization 209 (*see also* workforce)
productivity 9, 47–9, 55, 57
Lakeside, Thurrock 8, 80, 81, 220
legislation 77, 83–5, 104, 149, 165, 166, 179, 193, 194, 206, 207, 209, 227, 228, 230, 246, 251
leisure
activity 8, 154, 197, 204, 206, 208, 210, 211, 212, 214, 219, 220, 222, 227, 228, 257
shopping 13, 208–28, 257
time 4, 187, 188, 195, 204, 209–12, 215, 224, 226, 227, 257
leveraged buyouts (LBO) 50, 58
Liberty 2, Romford 218, 220
limited-line discounters 9, 42, 53, 54, 61, 65–7, 190
local government support 103
Loi Royer 30, 85, 149

Mall of America, Minneapolis 225
malls 4, 8, 79, 82, 88, 98, 105, 147, 148, 149, 155, 156, 160, 161–3, 166, 168, 219, 221, 225, 227
management
expertise 9, 15, 16, 36
information 5, 38
managers 13, 17, 39, 64, 197, 208, 221, 224–8, 231, 247
manufacturer–retailer relations 45
market niche 53, 119, 238, 249
market segmentation 5, 216
marketing 4, 5, 21, 23, 24, 31, 72, 115, 229, 234, 238, 248
Martin Luther King Jr Center, Los Angeles 162
Meadowhall, Sheffield 8, 11, 80, 110, 118, 119, 121, 143, 144, 220
Merry Hill, Dudley 8, 121, 144
MetroCentre, Gateshead 8, 80, 84, 121, 141–3, 145, 219
models 25, 38, 77, 86, 112, 152, 214, 228
Monopolies and Mergers Commission (MMC) 48, 52
multi-storey car parks 12, 105, 106, 139, 146, 165, 166, 254
multiplier effect 154, 156

natural surveillance 157, 168, 169
neighbourhood effect 130, 139
neighbourhoods 152, 190

Office of Fair Trading (OFT) 46, 48, 52, 54, 63, 64
oligopsonistic buying power 41, 45, 48, 54
out-of-centre 7, 8, 10, 11, 49, 86, 110, 112–15, 116, 117, 121–6, 128, 132, 138–40, 147, 150, 151, 173, 241, 252
out-of-town 8, 10, 71, 76, 79, 82, 85, 86, 104, 107, 141, 145–7, 154, 155, 207
outward movement of retail capital 42, 66
own-label trading 45, 59

packaging 4, 115, 237, 245
park-and-ride schemes 146
Parly 2, Paris 84
parques commerciales 72
part-time employment 3
patterns of consumption 4, 115, 243, 248
pedestrian
flows 141, 143
subways 12, 165, 167
trips 161
pedestrianization 105, 146
pedestrians 160, 162, 166–8, 174, 255
Piece Hall, Halifax 223
planned shopping centres 10, 12, 70–74, 75, 78, 80–82, 86, 87, 98, 149, 219, 223, 250, 255
planners 7, 13, 78, 126, 146, 156, 160, 173, 179, 181, 186, 224, 254, 256, 258
planning constraints 4, 6, 10, 44, 251

Planning Policy Guidance 122, 146
population
  ageing 3
  distribution 3
potential mobility 174, 190
Prestel 183–8
price competition 51
profit margins 9, 44, 47–9, 51, 52, 53, 54, 55, 57, 58, 62, 67, 249
public transport 12, 133, 143, 146, 148, 149, 162, 164, 166, 168, 172–5, 177, 180, 182, 190, 250, 253, 256, 258

recession 6, 50, 51, 53, 55, 59, 101, 121, 123, 144, 219, 221, 230
redevelopment 10, 12, 85, 98, 101, 104, 121, 122, 145, 146, 157, 163, 253
regional shopping centres 7, 8, 11, 73, 79, 81, 84, 98, 116, 125, 140, 141, 144, 146–50, 251, 252
regulatory environment 9, 43, 48, 51, 52, 57, 58, 249
rents 98, 116, 219
retail, retailing
  change 2–6, 8, 9, 11, 12, 14, 88, 107, 109, 110, 131–3, 141, 145, 152, 171, 247, 249, 250, 252, 258
  cross-border 10, 16, 46, 84
  concentration 5, 9, 41, 248, 249
  corporations 5, 41, 45, 47, 176
  distribution 2, 10, 26, 82, 182
  hierarchy 6, 8, 80, 121, 125, 141, 175, 190
  innovation 7, 249, 252, 253
  parks 10, 11, 81, 116, 121, 136, 137, 141
  planning 8, 134, 141, 149, 190, 248, 252, 253, 257
    policy 252, 253
  price maintenance 5
  revolution 8, 10, 12, 151, 172, 258
  warehouse park 71, 79, 134, 135, 144
  warehouses 7, 8, 134–8, 140, 150
Retail World, Gateshead 135
retailer–supplier relations 41, 44
revitalization 77, 124, 147–9, 151, 161, 254
Rittenhouse Square, Philadelphia 157
Ruhrpark 73
rural populations 175
Rush Street, Chicago 160, 168

Safer Cities Programme 155
safety 11, 48, 123, 153, 155, 156, 161, 162, 164–9, 255, 258
St David's Centre, Cardiff 100, 101, 102, 104
Saturday working 193, 202–204, 209, 221
scrambled merchandising 5
"sheddies" 4
shift work 195
shopper
  boredom 208
  safety 11, 258
  security 11, 153, 162, 168, 254
shopping
  behaviour 6, 11, 12, 126, 130, 131, 133, 136, 141, 150, 151, 211–15, 217, 218, 220, 224, 241, 247, 257
  centres 7, 8, 10, 11, 12, 13, 17, 70–74, 75, 77–82, 84–7, 91, 98, 102, 110, 114, 117, 116, 125, 127, 130, 131, 132–5, 140, 141, 142, 144, 145–50, 154, 162, 168, 172, 174, 177, 181, 212, 214, 216, 217, 218, 219, 221, 223, 225, 227, 228, 249–52, 254, 255, 257
  malls 88, 89, 105, 147, 155, 160, 227
  top-up 133
Shops Acts (1928, 1936, 1950) 193, 205
Single Market 19, 27, 195
16th Street Mall, Denver 161–3, 166, 168
sourcing 9, 15–22, 25, 26, 28, 40, 248
spatial switching 41, 60, 66
speciality centres 10, 77, 81, 82
store-expansion programmes 44, 49–51, 59, 60
suburbanization 6, 88, 111, 251
subways 12, 165, 167, 177
Sunday working 13, 52, 193–7, 202–207, 214–16, 226–8
Sunday trading 13, 193, 196, 206, 226, 228
supermarkets 18, 35–7, 52, 58, 59, 63, 66, 79, 96, 127, 132, 172, 173, 176, 178, 184 185, 211, 213, 217, 236, 237, 243
superstores 4, 7, 8, 11, 12, 49, 50, 54, 64–7, 96, 115, 121, 126–34, 134, 135, 137, 138, 139, 140, 145, 150, 151, 172, 173, 174,

176, 178–80, 182, 189, 252
saturation 134
Swansea Enterprise Zone 7, 128, 135, 136, 140

technology 5, 9, 12, 20, 26, 29, 31, 37, 47, 174, 182, 186, 187, 189, 240, 249, 258
television programming 182
theme shopping 221
Town and Country Planning Act 181, 187
town centre 77, 118, 122, 127, 135, 142, 144–8, 160, 180, 218
trading hours 127, 145, 193, 194, 196, 205, 226
traditional centres 134, 138–40, 145–7, 150, 151, 173, 251–4
traffic calming 124, 146
traffic congestion 11, 88, 133, 141, 146, 151, 252, 253
Trafford Park, Manchester 123
transfer of management expertise 9, 15, 16
trips 2, 126, 127, 129, 132, 133, 138, 151, 161, 166, 173, 176, 177, 181, 191, 204, 212, 214, 243, 250
Trump Tower, Manhattan 162
typology, of shopping centres 30, 71, 80, 81, 86, 87, 250

unemployed 4, 111, 173, 192
Urban Programme 160, 164

Victoria Centre, Nottingham 164
videotext 182, 183
viewdata systems 5, 182
village stores 175
Ville 2, Charleroi, Belgium 85

Water Tower Place, Chicago 162
weekend market 223, 224
West Edmonton Mall 79, 149, 219, 225
White Rose Centre, Leeds 123
women 4, 12, 13, 155, 156, 158, 164–7, 169, 172–4, 176, 177, 180, 181, 192, 193, 195, 197–202, 206, 207, 256
workforce 2, 4, 13, 43, 103, 192, 195–7, 205, 206, 249, 256
working hours 13, 192, 195, 198, 205, 209
non-standard 196, 198–204, 207